THE CITY

URBAN FOCUS

URBAN PLANNING

URBAN FOCUS

MARK HEWLETT

Routledge
Taylor & Francis Group
LONDON AND NEW YORK

First published in 1974

This edition published in 2007
Routledge
2 Park Square, Milton Park, Abingdon, Oxon, OX14 4RN
Simultaneously published in the USA and Canada by Routledge
711 Third Avenue, New York, NY 10017
Routledge is an imprint of Taylor & Francis Group, an informa business

Transferred to Digital Printing 2007

First issued in paperback 2013

British Library Cataloguing in Publication Data
A CIP catalogue record for this book
is available from the British Library

Urban Focus

ISBN13: 978-0-415-41740-2 (volume hbk)
ISBN13: 978-0-415-41935-2 (subset)
ISBN13: 978-0-415-41318-3 (set)
ISBN13: 978-0-415-85185-5 (volume pbk)

Routledge Library Editions: The City

General Studies

Mark Hewlett

URBAN FOCUS

METHUEN EDUCATIONAL LTD
London · Toronto · Sydney · Wellington

First published
by Methuen Educational Ltd
11 New Fetter Lane London EC4P 4EE

ISBN 0 423 88440 9 net
0 423 88430 1 non net

Printed in Great Britain
by Butler & Tanner Ltd, Frome and London

Contents

Objectives of urban study

1

Relevance, generality, disciplinary identification and the scientific approach.

As with any worthwhile enquiry, urban study aims to focus attention on important questions, to develop understanding of the circumstances in which they arise and, as far as possible, to suggest either the answers to the questions, or ways in which the answers may be arrived at.

The last decade has witnessed a considerable increase of interest in the urban environment. This has occurred as part of a trend towards application of intellectual disciplines to subjects of practical concern, associated with an erosion of the traditional distinction between the academic and the intellectual on the one hand, and the practical and socially relevant on the other. Perhaps more clearly than ever before, it is seen that the most socially relevant enquiry involves the adoption of certain critical objective attitudes to investigation, and the analysis of the modes of thought and conceptual structures of those academic disciplines which provide some of the keys to sorting out complex social and environmental problems.

The relevance of urban study in terms of its relationship to important questions seems obvious enough in a country where, in addition to the 80 per cent who live in towns and cities, the remaining 20 per cent are also part of the urban system. Not surprisingly such a study touches on a wide range of issues which affect our daily lives. This alone justifies this book, but it is also realized that if we do not face up to many of these issues we run the risk of jeopardizing the quality of our lives in the future. While *Urban Focus* is by no means a further addition to the environmental doomsday catalogue, no apology is felt necessary for putting forward as a major objective the development both of a sense of responsibility for our social environment and a sense of commitment to doing something about it. Social and environmental problems are alluded to throughout the book. Three sections are particularly concerned with these problems. They are sections 9, 10 and 11.

Urban Focus, however, is devoted not so much to this task of persuading people that there are problems which demand our urgent attention, but to presenting a variety of questions and problems within a context of general ideas and information which illuminate the nature of the urban phenomenon. While reference is made to concrete examples, even in the most theoretical sections, the book is designed essentially as an abstract, setting out a general frame of reference consisting of:

(a) general concepts and theories
(b) methods of enquiry and research
(c) different ways of thinking about environment
(d) different types of data and how they may be examined.

The criterion of generality has determined the selection of material and its presentation, the aim being to equip the student with a set of ideas which can be transferred to the analysis of any urban environment. This generality is evident in the abstract and conceptual nature of some of the material, in the relationship of the work to wide areas and large numbers of people, and furthermore in the introduction of concepts and methods which may be taken up by the student in other areas of enquiry, particularly those which, like urban study, cross traditional disciplinary boundaries.

It is however necessary, if the work is to be rendered vital and realistic and if superficiality is to be avoided, that generalizations are tested against the careful observation and analysis of real situations in a detailed case-study of some particular area, preferably the local area with which the student is already to some degree familiar. Instructions are provided to enable a local case-study to be undertaken in the most profitable way.

Implicitly through the case-study, and elsewhere more explicitly, there is a strong emphasis on the learning process which it is believed can best be promoted by providing guide-lines for individual or group work and by posing questions which demand an active response on the part of the student. Although this book includes its own information and ideas, further sources of information should be obtained. In addition to the actual environment of places and people, these are of two main types: (a) books and other secondary sources of information deliberately given a prominent place in a separate section and selected to include sources which are easily available: (b) other students whose views will be invaluable where discussion is required prior to the formulation of balanced judgements in the face of a variety of opinion.

One reason for emphasizing the learning process and the desirability of further enquiry on the part of the student is an awareness – inevitable where such a large subject is dealt with in such a short book – of what is omitted or treated very briefly, e.g. the organization of local government, problems associated with the urban explosion of the Third World, a more detailed and explicit treatment of urban history.

Decisions about what to emphasize and what to leave out have been made consistent with the criteria of selection already mentioned: relevance and generality. This is reflected in the title of the book, which connotes two essential characteristics and objectives: (a) the presentation of the urban environment as a system in which the human activity of areas focuses on central places, (b) the use of urban study as a means of bringing together the ideas of different disciplines to focus on to one subject.

Both these objectives are closely associated with the idea of integration and with the inter-disciplinary approach. However the book is equally if not more concerned with identifying and separating the different disciplinary elements and is thus more multi-disciplinary than inter-disciplinary. Each discipline provides its own structures, concepts and perceptions which constitute important learning mechanisms which assist in the task of clarifying, simplifying and making sense of the mass of heterogenous data that makes up our complex and potentially confusing social environment. The disciplines concerned are geography, economics, demography and sociology.

Finally, to anticipate the response-demanding nature of the book, the reader is recommended to consider what the objectives of urban study should be and to criticize the book in the light of any conclusions reached.

2 Understanding the urban environment: three analogies

The attempt to find an **analogy** to represent all towns and cities may seem doomed to failure in the light of the enormous variety of them which exist. If, out of a list of towns and cities, one selects Wigan and Las Vegas, Grimsby and Jerusalem, Basingstoke and Baghdad, the points of similarity are less likely to spring to mind than the points of dissimilarity. Nevertheless it is maintained that different towns and cities are merely examples of different species (e.g. port, market town, capital) of the same genus (urbs), and that analogies can be drawn to illustrate the genus, i.e. to show the essential general nature of all towns and cities.

Before considering these analogies it must be emphasized that: (a) the town is seen as the central focus or nucleus of a surrounding area or **field** (b) an analogy is not the same as a **model** which is characterized by a greater degree of abstraction and mathematical precision (c) it is as important to examine the limitations of an analogy as it is to note its accordance with reality.

1 Mass and gravity

In order to explain the concentration of human populations, H. C. Carey reasoned in 1858 that social and physical phenomena are based on the same fundamental law. 'Man, the molecule of society is the subject of the Social Science. . . . The greater the number collected in a given space the greater is the attractive force that is there exerted. . . . Gravitation is here, as everywhere, in the direct ratio of the mass and the inverse one of distance.' (See the photograph on page 64.)

2 Factory

This analogy attempts to illustrate the internal arrangement of the town as well as its external relationships. The town is compared to the factory as a unit of functionally related parts with an inflow and outflow of materials along lines of communication linking it to the catchment area of its raw materials/components and its marketing and sales region. The division of labour in the factory, reflected in its layout, bears close resemblance to the variety of functions and their arrangement in the town. The reader is recommended to draw this analogy to include, among other components, the manager's office, the shop floor, the canteen, and storage and waste disposal facilities.

3 Organism

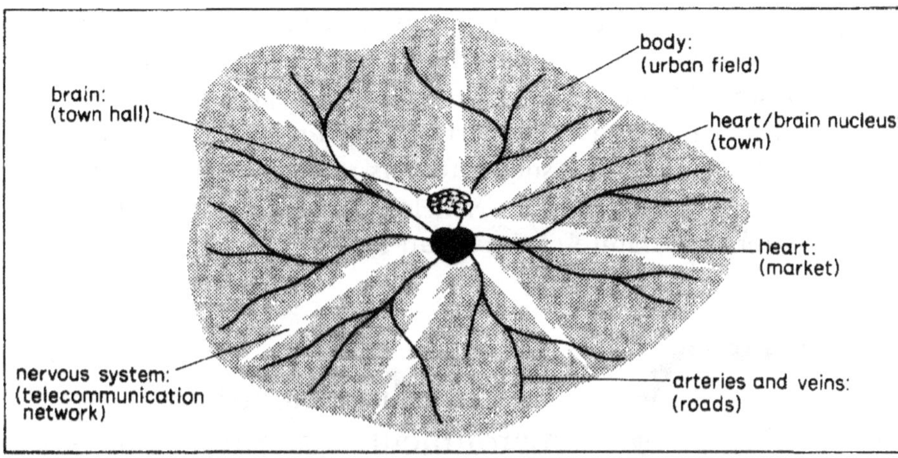

Fig. 1 Naïve representation of the urban system

This analogy likens the body to the whole urban field, that area surrounding the town which supports and in turn is served by the heart/brain nucleus which represents the town. The essential unity of body (**urban field**) and heart/brain nucleus (**town**) makes this analogy particularly useful.

The analogy makes use of the idea of circulation, in the organic sense of blood through veins and arteries pumped through the heart, and circulation in the economic sense of goods transported along the transport network of roads, railways etc. In this respect it is interesting to note the French for traffic (*la circulation*) and the use of the term **arterial** roads. Economically the heart of the system is the market place, the focal point of the transport network along which goods and services, the 'lifeblood' or the urban region, are transmitted. The inability of transport systems to cope in the course of time with the required flow of goods may be compared with increasing blood pressure, traffic jams with thrombosis and urban reconstruction with various surgical operations, but the analogy begins to lose its validity at this level of detail.

If the market place is the heart, the communal brain may be considered to reside in the Town Hall and Council Offices whose messages are transmitted along the nervous system representing the telecommunications and postal network; but the telephone exchange and post office are omitted, as is the means whereby the cells of the organism (people) communicate directly with each other.

Any attempt to add an analogy for the digestive and waste disposal systems of the body will tend to conflict with the blood stream analogy. A. J. Rose has drawn up a scheme of the urban metabolism which deals with these aspects separately.

The reader may draw comparisons from the lungs and other parts of the body. While such attempts tend to produce increasingly crude analogies they may nevertheless provide further valuable insights into the nature of the urban system.

3 Understanding the urban environment: a summary of major concepts and questions from contributory disciplines

Why towns and cities? Economic aspects of urban concentration

'In every part of the world where certain social and technological conditions exist populations have drawn themselves together to form towns and cities.' Most societies 'have found the magnetism of urbanism irresistible'. Today the world picture shows urbanization proceeding at an ever-increasing rate.

What are the reasons for the apparently inexorable growth of towns and cities?

✠'their rapid growth and large size testify to their superiority as a technique for exploiting resources'. (C. Harris and E. Ullman)

towns and cities owe their existence to the presence of certain social and economic activities which require concentrations of people, buildings and machines within relatively small areas'. (J. H. Johnson)

'towns and cities exist because they are more convenient than other forms of settlement. This convenience is common to virtually the whole range of social activities outside agriculture but is particularly apparent in the case of commerce and manufacturing.' (A. J. Rose)

Towns may be seen as socio-economic machines successfully designed for promoting the conduct of human affairs which involve commercial and social transactions, co-operative enterprises, the receiving and distributing of information and materials, movement to and from work, leisure, etc.

The common characteristics of these activities relevant to the spatial arrangement of people and buildings are **communication, transfer** and **movement** of goods, information and people. All movement involves the expenditure of energy/time/money which are scarce. The simple advantage of concentration is the reduction of this expenditure to a minimum. Although it has a specific connotation for the human geographer, Zipf's idea of human behaviour and the **law of least effort** is valuable in suggesting that things are arranged in the most efficient way to cut down the loss of unnecessary energy. Related to this is the concept of minimizing the **friction of distance** which gives rise to a number of **centralistic principles**, one of which is that 'human occupance is focal in character' (Garner).

The logic of commercial concentration

Stage 1 Consider a pre-urban agrarian society of independent **subsistent** producers dispersed more or less evenly so that each producer has sufficient land for his own requirements. No **surplus** is produced.

Stage 2 Should two producers come to realize that they do not produce each of their various requirements with the same expenditure of effort, they may decide to **specialize** in whatever they can produce most efficiently and make up their respective deficiencies by **exchanging** or **trading** their **surplus** by means of a **barter** agreement.

Stage 3 Assume that this idea spreads and that all the producers in an area specialize in different products and that barter agreements are made between each of the producers. Each transaction involves a separate journey at a different time. This results in a transport network which links each producer to all other producers.

Stage 4 If all the producers arrange to exchange their surplus at a particular time in a particular place, there are considerable savings in time and distance. The place selected for the exchange becomes what is known geographically as a **Central Place**: a place where a service – or more usually a range of services – is made available to the population of a surrounding area. The Central Place need not be geometrically central.

Stage 5 Once established as a commercial centre, in addition to those who may in the course of time specialize in commercial affairs, others (e.g. inn-keepers, craftsmen) will tend to **agglomerate.**

What factors will promote or retard the development of a commercial Central Place? At what stage of development will the Central Place become a town?

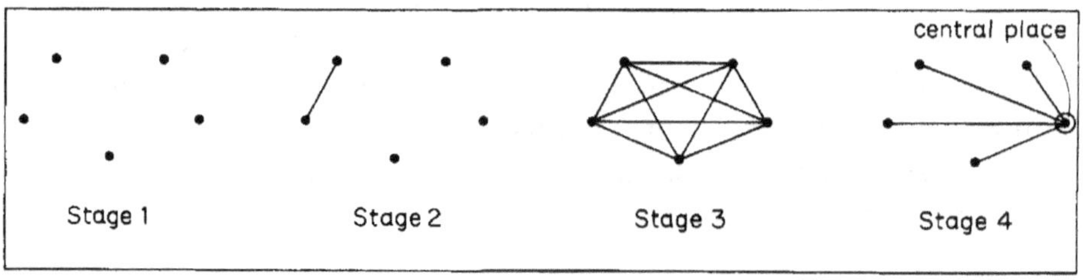

Fig. 2 Transport links between producers at Stages 1–4

What will be the saving in total distance travelled between Stage 3 and Stage 4? What would be the corresponding saving if the exchange arrangements involved 4, 8, 12, x producers?

The logic of industrial concentration

Consider circumstances of technological advance, where an independent producer (see Stage 1 above), designs and manufactures an implement which enables him to save time and/or increase his **output/productivity**, and which prompts a decision to concentrate his efforts in producing more of them to sell in exchange for his other requirements.

The links required for the resulting trade in implements will establish the manufacturer as a focus of transport, and in as far as services are being provided for a surrounding area a Central Place will have come into being.

Should the demand for implements increase, the manufacturer may be motivated to increase production and employ assistants who will tend to reside at or near the factory; he may require men to look after the store-sheds, book-keepers to maintain records of transactions etc. As in Stage 5, see page 14, in addition to those involved in the industrial activity, others will tend to agglomerate (e.g. bankers, specialist teachers to cope with the demand for people able to read, write and master technical skills).

Today towns and cities are not always the best places for industry but in seeking a **least cost location** where costs are minimized, towns offer the following advantages:

1 lower **distribution** costs (proximity of a large 'market' for sale of produce)
2 lower costs of **assembling** material from other industries in the town.
3 a large **labour** supply
4 established **infrastructure**: facilities such as water supply, gas, public transport provided by municipal authorities
5 **transport** facilities.

The term **external economies** is given to the saving on costs as a result of conditions outside the factory.

What other advantages are there for industry in towns?

What disadvantages are there for industry in towns?

How close is the relationship between industrialization and urbanization?

What were the special reasons for the growth of towns in particular places during the industrial revolution?

The logic of political and social concentration

While all aspects of human agglomeration cannot be reduced to economics (at least as it is normally understood), the efficient provision of government, administration and defence, as well as other aspects of social life, such as public entertainment, need to be considered in terms of least cost location.

The concentration of professional and administrative functions in towns and cities is important not only for the concentration of those directly concerned with them but also for their effect in attracting other supporting services and population. A. J. Rose asserted that 'the political nucleus has been the most potent generator of towns and cities'. In considering also the concentration of other types of social provision (opera houses, theatres, football stadia) two basic questions arise:

Would it make very much difference if they were situated elsewhere?

What effect do these functions have in promoting further concentration?

The urban nucleus and its surrounding field: the external geography of towns and cities

Terminology

The processes which operate to focus activity in Central Places have been explained in terms of functional efficiency. The term **urban functional region** therefore appropriately describes the area which consists of the urban **nucleus** and its surrounding **field** which it serves and with which it is functionally related.

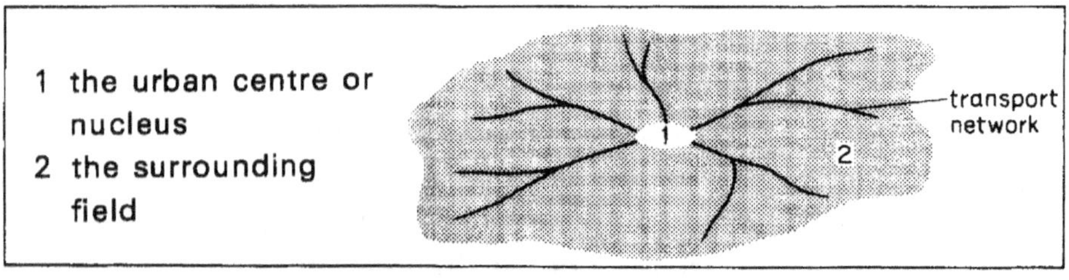

Fig. 3 The urban functional region

Some alternative equivalent terms:

urban region: functional region; centred region; city region
urban nucleus: node; pole; focus; Central Place
urban field: umland; zone or sphere of influence; hinterland; catchment area.

Land value and the arrangement of land use in the urban region
Because of advantages to be gained from a central accessible position in the urban region, the value of land will be greatest at the centre and diminish with distance from it.

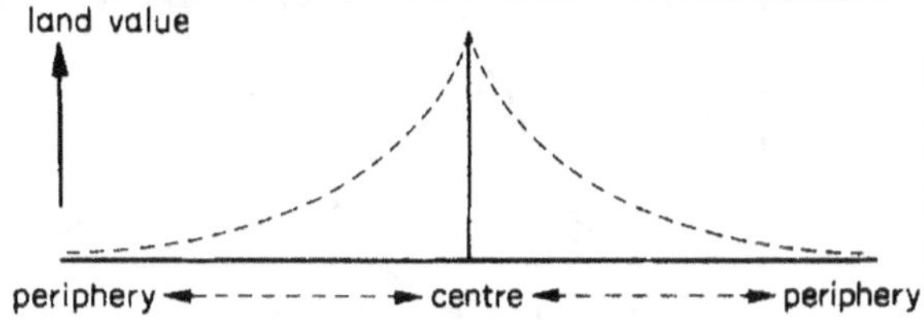

Fig. 4 Land value gradient in the urban region

Competition for the most valuable land at the centre will force up its price. Only those who can make intensive use of the land can afford this. Throughout the urban region there is a direct relationship between proximity to the centre, land value and price, and land use intensity. One would expect the gradation in land value to be reflected in a gradation of land use in the form of concentric zones.

1 very intensive land use: continuously built-up area
2 and 3 moderately intensive land use: market gardening, dairy farming, requiring access to the city for daily delivery
4 extensive land use: arable and other farming

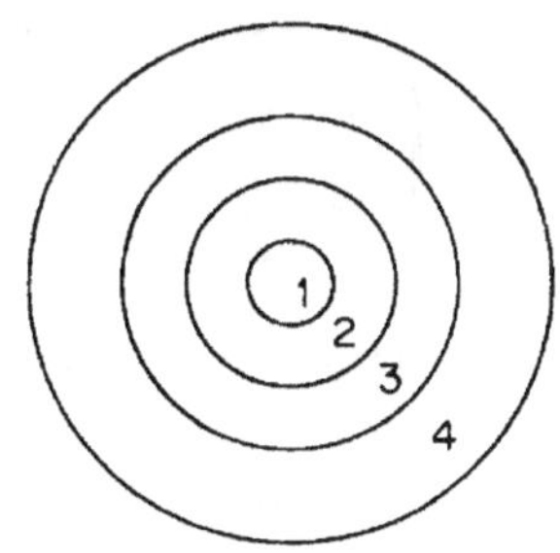

Fig. 5 Concentric land use zones in the urban region (after von Thunen)

What factors disrupt this theoretical arrangement?
How far is this pattern observable in reality?

An environment of urban regions: the idea of a hierarchy of settlements and functions
On a uniform surface the urban region will be circular. Why? But

if the theory is extended to include more than one region there will either be gaps between regions or overlapping. Elementary **packing theory** shows that hexagonal forms will develop.

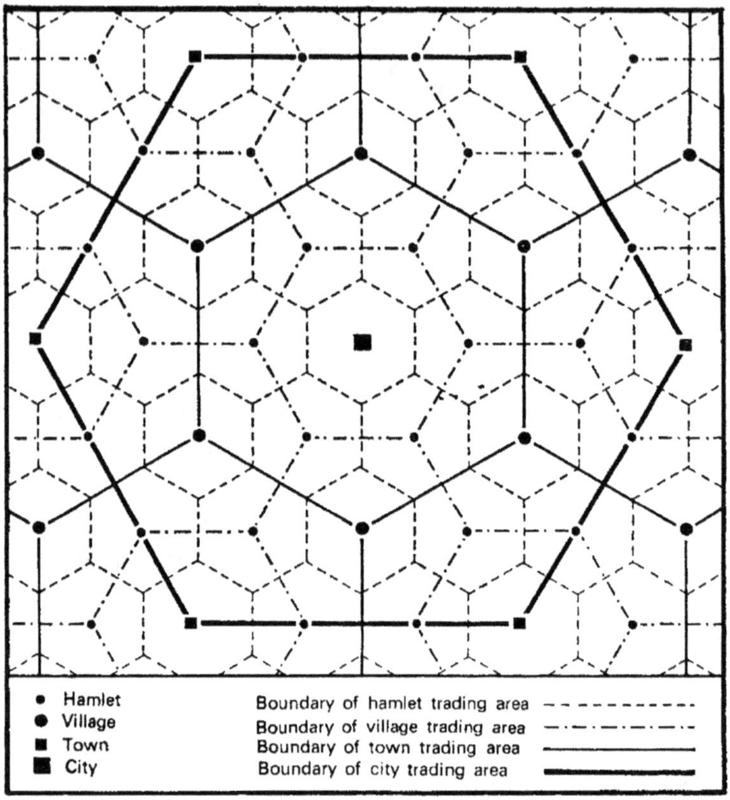

Fig. 6 The arrangement of trade centres (Christaller's theory)

In drawing up the theoretical urban arrangement, see above, Christaller proposed that towns with the lowest level of specialization would be equally spaced and surrounded by hexagonal hinterlands. For a given number of these centres there would be a more specialized **higher rank** centre which in turn would be situated at an equal distance from other centres with the same level of specialization. In this system there is a constant ratio between the numbers of trade centres at various levels of specialization and between the areas of their respective zones of influence.

This theoretical arrangement has been applied to settlements as administrative and cultural centres. For further explanation of Central Place theory the reader is recommended to refer to *Urban*

Geography by J. H. Johnson (Pergamon, 1967) and *Settlement Patterns* by J. A. Everson and B. P. FitzGerald (Longman 1969).

Settlements and their spheres of influence differ in size. Some towns (e.g. local market towns) serve only limited areas, others (e.g. county towns) serve more extensive areas, which include within them several small towns and their local spheres of influence.

The concepts of urban rank, high- and low-order services and threshold populations.

All settlements – from village to city – provide certain basic services and goods (known as **convenience goods**) required on a day-to-day basis. Such goods and services have a low **population threshold**, i.e. they do not require a large population to support them. As they are required relatively frequently, and as the goods in this category are usually inexpensive, people are unwilling to travel far to obtain them: they are said to have a small **range**. Examples of such goods and services are post office, public house, newsagent and tobacconist.

Above this basic layer of goods and services some more specialized higher order goods and services occur less frequently. Goods in this category are more expensive, required less often, have a greater range and a higher population threshold. The presence of these types of service and goods are indicative of a higher **rank** of settlement.

	general store	public house	post office	secondary school	clothes shop	bank	cathedral	university
village	✓	✓	✓					
town	✓	✓	✓	✓	✓	✓		
city	✓	✓	✓	✓	✓	✓	✓	✓

✓ function appears in that type of settlement

Fig. 7 Examples of 'indicator functions' of different types of settlements, as suggested by J. A. Everson and B. P. FitzGerald in *Settlement Patterns*, Longman, 1969.

N.B. The importance of a centre may be quantified by allocating numerical values to each example of a service observed in a centre, see section 13, page 98. For a more elaborate classification of settlements, see section 6, page 39.

 The urban region in reality

Section 13, page 97, indicates ways in which the topic may best be studied.

The map of Wrexham's urban field suggests that the idea of the urban region is less simple than it appears in theory as the zones of influence for different functions do not coincide.

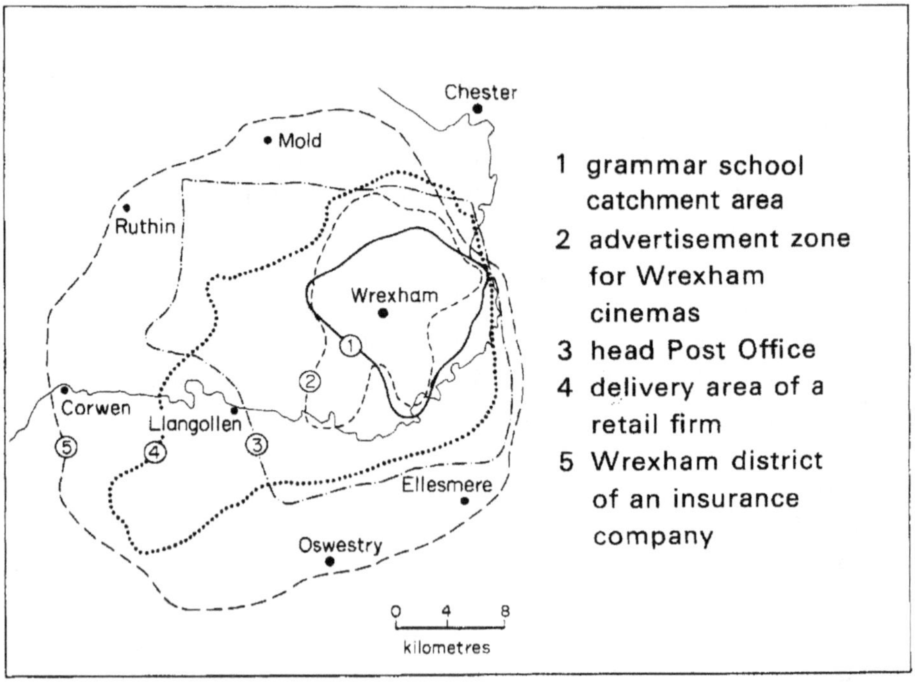

Fig. 8 Wrexham: zones of influence (after H. Carter, see *The Towns of Wales*, H. Carter, University of Wales Press, 1965)

Urban components and their arrangement: the internal geography of towns and cities

The material structure of the town consists of buildings and spaces which enable the town to fulfil its various functions. A list of the main urban components is given in section 13, page 93.

The main purpose of this section is to convey important general ideas about the structure or **morphology** of towns and cities and will be most useful if the reader draws comparisons with his or her own town, a map of which should be available for reference.

Functional zones

The buildings etc. associated with the town's various functions are not spread randomly; they cluster into areas known as **functional zones**, where particular types of land use predominate. These are more or less clearly segregated.

While no two towns are exactly the same, certain patterns of land use recur with sufficient regularity for it to be possible to construct models or generalized schemes of urban land use.

1 Central Business District
2 zone in transition
3 zone of predominantly working-class housing
4 zone of predominantly middle-class housing
5 commuter zone

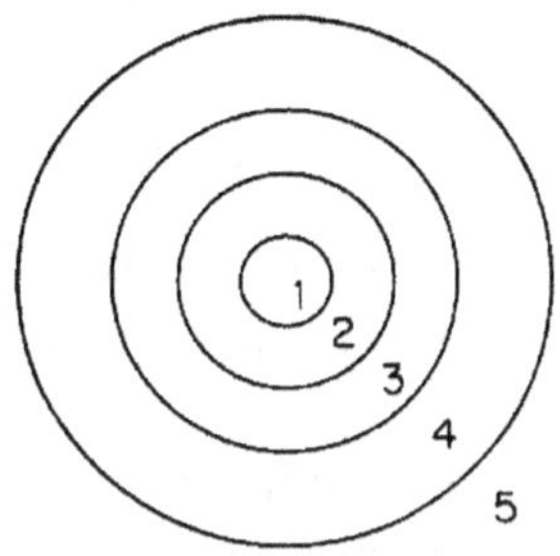

Fig. 9 The concentric pattern of urban land use (after E. W. Burgess's theory)

N.B. Commuters – those who travel some distance to work – may live in zone 4. Most are assumed to commute to the CBD.

1 *The Central Business District (CBD)* is the commercial, civic and social core not just of the town but of the whole urban region, being at the focus of the regional transport network. The main railway and bus station(s) are to be found in or at the edge of this zone. Today the zone is often surrounded by an inner ring road to relieve the congestion which would otherwise be intolerable.

Here land values and land use intensity are at their highest as firms compete for the best sites. In order to maximize use of expensive land, high buildings are constructed. (Why are not all buildings high?)

Characteristic buildings of this zone are major office buildings, especially headquarters of firms, large specialist shops, municipal buildings, theatres and major entertainment facilities. The CBD is invariably subdivided into commercial, administrative and professional zones. It should be noted that not all functions found in the CBD serve the whole urban region.

2 *The zone in transition.* This zone, which has three main elements, includes some of the worst elements of urban blight and is some-

 times termed the *zone of discard* or *decay* and/or *twilight zone*, where land uses and functions are obsolete and in decay.

(a) **the wholesale business district** the 'storehouse' of supplies for the CBD
(b) **old small-scale industry** established when the town was less extensive
(c) **poor housing, now falling into disrepair slum** housing, often accommodating immigrant groups in what may be loosely termed **ghetto** concentrations, and often with high crime and poverty rates.

The whole of this zone may be considered physically and socially transitional and unstable. Today it is in this zone that major attempts at **urban renewal** are taking place to alleviate sub-standard conditions. As a result, the combination of old housing, new high-rise flats, and derelict waste land awaiting development is particularly characteristic.

3 *Zone of working-class housing* inhabited by workers who remain nears established areas of industrial employment but who have 'escaped' from the zone in transition.

4 *Zone of predominantly middle-class housing* characterized by a higher proportion of single-family dwellings, in better physical conditions and built at lower densities. In addition to private middle-class suburbia, council estates have also been built in this zone. Both types of housing have contributed to urban sprawl which has often swallowed up small villages once beyond the town boundary. In many cases these villages have become the focus of local (**neighbourhood**) shopping and social facilities.

5 *The commuter zone*. Beyond the town/city boundary but functionally part of it, the zone contains detached middle-class suburbs (**exurbs**) in the form of commuter villages (around towns) or commuter towns (around cities). The commuter zone may be separated from the city by a **Green Belt**, where building is restricted.

This alternative simple model retains the CBD but replaces the concentric zones with sectors which develop outwards from the functional differences that appear in the inner part of the town. As the town expands, the differences are perpetuated, as new buildings of similar function or type are constructed adjacent to the old, thus forming sectors or wedges which develop as suggested by the arrows.

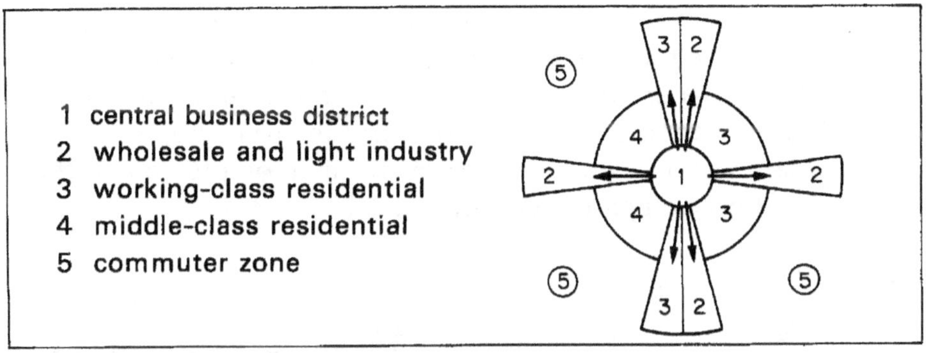

Fig. 10 The sector pattern of urban land use (after H. Hoyt's theory)

N.B. The equivalent of the Green Belt is the green wedge.
The zone in transition encircles the Central Business District.

Which of these two patterns appears to fit best with the arrangement of towns or cities with which you are familiar?

A third variant is the **multiple nuclei pattern.** What are its characteristics? (See *Urban Geography*, J. H. Johnson)

The hypothetical British city

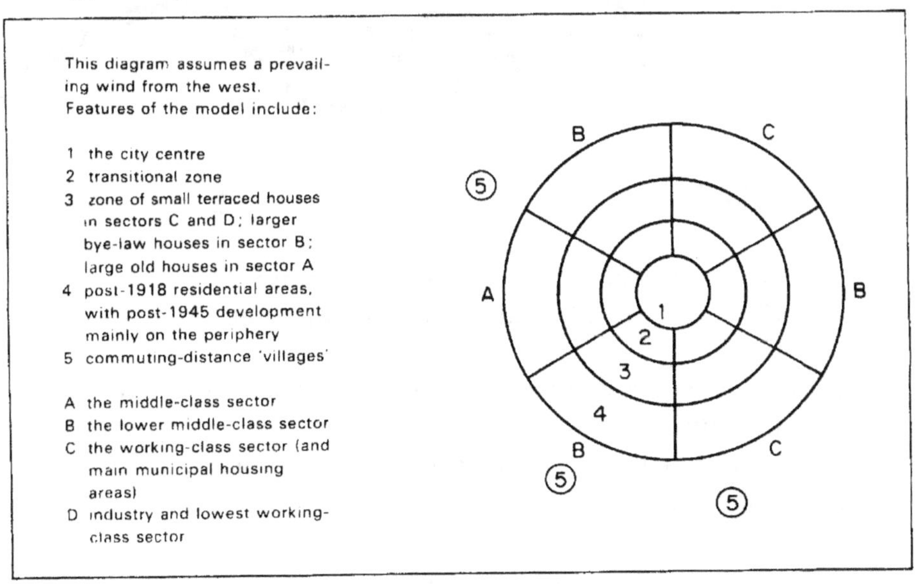

Fig. 11 The structure of a hypothetical British city (see *Approach to Urban Sociology*, P. Mann, 1965)

 In combining the concentric and sector theories, this model attempts to give the best generalization of the structure of a well-established town or city in Britain. To test the accuracy of this model against an actual example this grid (expanded to an appropriate scale) might be placed over a map of a town and the actual land use compared with the model.

A typical English country town

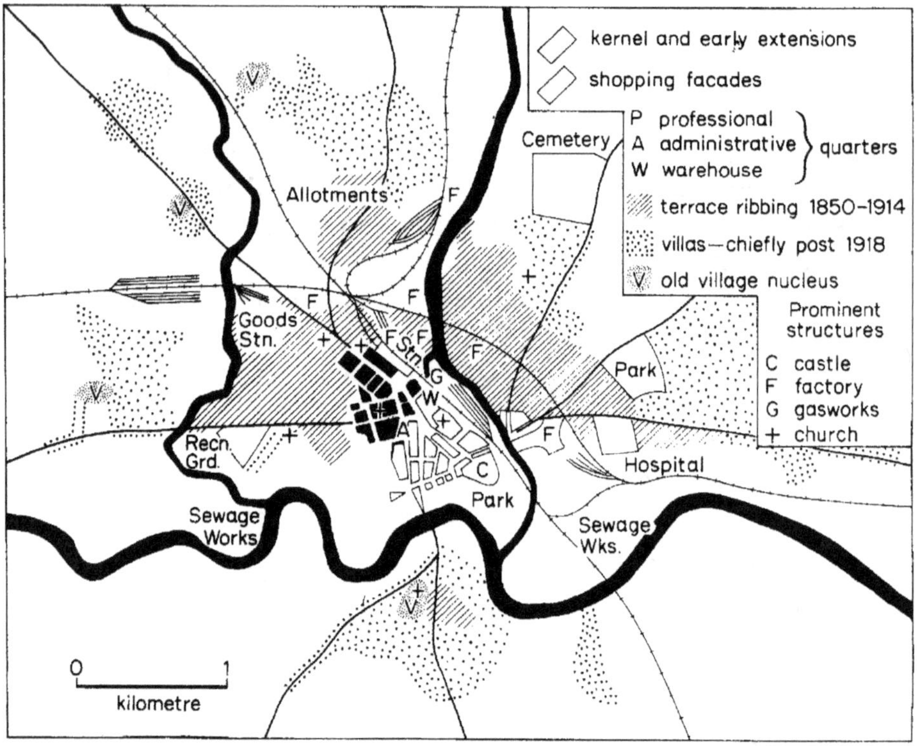

Fig. 12 Morphological diagram of a county town.
NB Villas – residential housing mainly detached.

This less generalized model may again be used to compare with an actual example, whose characteristics will be the result of a combination of factors peculiar to it, such as its site and its historical development.

The urban population: demographic aspects

For a general introduction to urban demography, Chapter 7 of

Towns and Cities by E. Jones (OUP, 1966) is particularly recommended.

1 *Urbanization*

Urbanization in its broadest sense is a process by which a population becomes involved in an integrated social and economic system organized around urban centres. While a population may become fully urbanized without everyone living in towns or cities, statistically urbanization usually refers to the proportion of a population living in urban centres or to the rise in this proportion.

Urbanization is closely linked to economic development. The following graph shows the cycle of urbanization through which countries pass as they change from agrarian to industrial society.

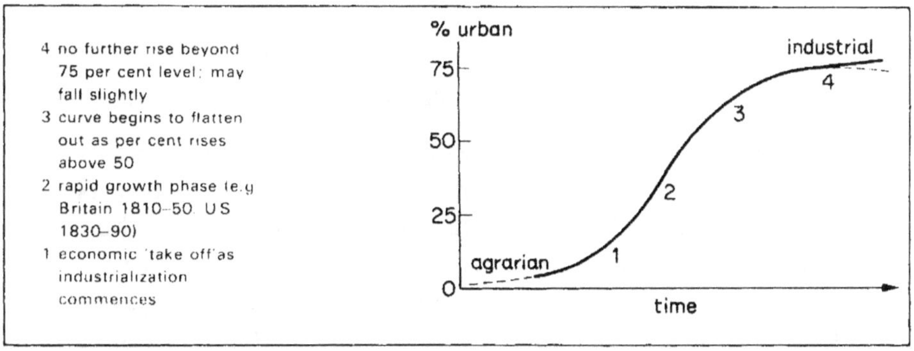

Fig. 13 The world urbanization curve (NB Urbanization is used here in its narrow sense meaning the proportion living in towns and cities)

Why do urbanization figures differ so widely from one country to another?

2 *People and space: population density*

Examination of population densities within towns and cities reveals that certain regularities exist, such as the population gradient shown in the following graph which indicates the relationship between residential population and distance from the centre. It may be compared to a volcano subsiding and spreading outwards with time.

What would the working hour population gradient be?

What is the relationship between this graph and the graph of urban land value?

What factors influence population density?

Why does the 'volcano' subside with time? Is this trend likely to continue?

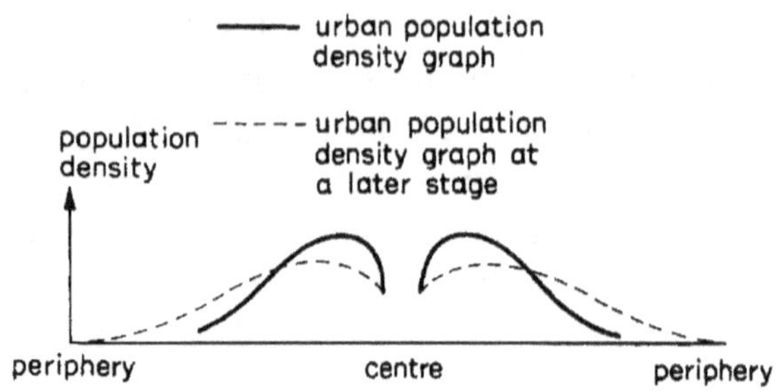

Fig. 14 Population density across the city

3 *Population migration*

Three types of migration are particularly relevant to urban study:

(a) Rural-urban migration has been a major factor in urban growth. A distinction is often made between **push** factors which are related to particular deficiencies of which the potential migrant in an area becomes aware, and **pull** factors which attract the potential migrant to a particular place. Garnier and Chabot refer to the 'magnetic attraction of towns', but they note that this rests as much on 'a psychological idea as on economic fact'. Hence the migration of people, in some cases, to towns and cities where conditions are found to be even worse than those they left behind.

(b) There is a constant movement of population between towns and cities, particularly in highly-developed industrial economies, mainly as a result of (i) variations in employment opportunities (this would include movement of individuals who are promoted to better jobs) (ii) movement to suitable places of retirement.

(c) Each day there is massive short-distance migration as people travel to and from work etc.

In what circumstances does rural-urban migration take place at a rapid rate? Which sections of the population are most mobile geographically? Could the massive daily movements of population profitably and practicably be reduced?

4 *Death rates*

In 1841 the average expectation of life in London was 36 years and in Manchester 26. Though these figures are extreme, death rates in many of the big cities of the nineteenth century were very high, a direct result of disease and ill-health arising from insanitary overcrowded conditions and perhaps from a sense of hopelessness amongst the many who led lives of privation and misery.

However, with improvements in sanitation, the readier availability of medical aid, better food, a higher level of (health) education, and reduction in overcrowding, urban death rates fell from a figure well above that for rural areas to one significantly below. In recent times there has been a tendency for urban and rural death rates to equalize as the extension and improvement of public health services have reduced urban-rural differences.

Within towns death rates vary significantly from one area to another: between, for example, new housing estates inhabited mainly by young married couples with children and older inner urban areas with a high proportion of older people.

There are also marked differences between towns, for example, between new towns with a high proportion of young workers and coastal resorts favoured by retired people.

5 *Birth rates*

It might be expected that the in-migration to towns and cities of young adults of reproductive age would result in higher birth rates in urban than in rural areas. In many cases this is so but equally interesting in its relationship to social and economic aspects is the fact that during periods of urban growth, birth rates in urban areas have been persistently lower than might have been expected.

Several reasons have been suggested:

(a) urban dwellers had stronger aspirations to a higher standard of living in a status-orientated society
(b) The cost of rearing children was much greater in a situation where a longer education was required for children to gain eventual access to more desirable and demanding occupations
(c) employment opportunities for women were greater in towns,

taking them out of the home, re-directing their energies and changing their attitudes. 'Thus absorbed and emancipated – the two things together – they develop a greater than normal tendency towards the practice of contraception'. (Mme. Beaujeu Garnier)

(d) knowledge of contraceptive techniques was greater amongst the more sophisticated urban population.

As in the case of death rates urban-rural differences have generally diminished, as urban values and standards have pervaded the whole of society.

Variations occur within towns and cities and between towns and cities: see the examples quoted above for death rates.

6 *Population pyramid: age composition and sex ratios*
The age composition of populations is obviously affected by birth and death rates and is referred to above. Sex ratios also show some significant variations, e.g. women invariably outnumber men in old populations and men tend to outnumber women in the rapidly growing cities of the developing world.

The age pyramid is a useful method of summarizing this data. (Note that age composition statistics are given in section 6, page 43, for London, new towns etc.) These could be translated into Population pyramids which would provide an interesting contrast

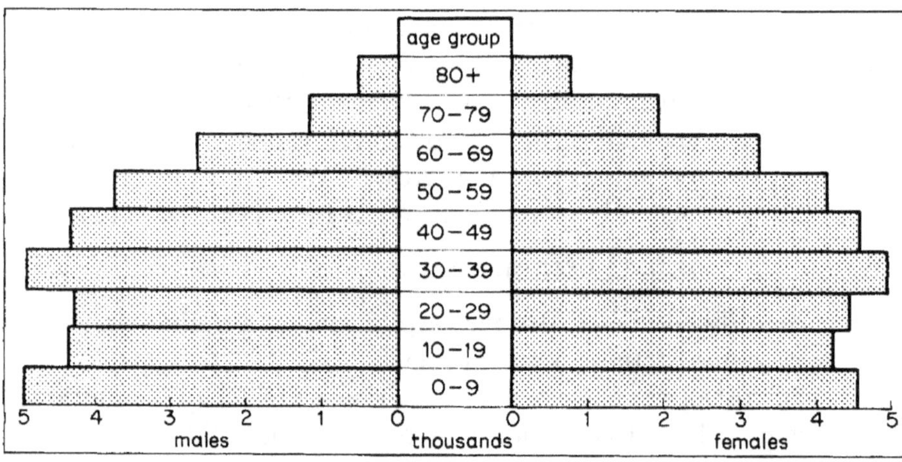

Fig. 15 Population pyramid of a typical medium-sized town.

to the one in Fig. 15, considered in *Town Study: A Sample Urban Geography*, by K. J. Driscoll, (George Philip, 1971), to be typical of a well-established medium-sized city or town (e.g. Bedford).

N.B.

1 the varying proportion of males and females in each age-group, especially in the first and last years.
2 the usefulness of this type of diagram, especially if presented in a 'percentage' form, for comparing the demographic structure of two or more towns, cities, or of areas within towns/cities.

Urban society: sociological aspects

The **sociological** aspects of urban study (as distinct from **social** aspects which, broadly speaking, include everything in this book) are concerned with the scientific analysis of social groups and the relationship between factors that influence social behaviour.

The summary given here is restricted to a number of major concepts which it is recommended should be examined further by reference to (i) a sociology textbook to help in clarifying some of the terms used (ii) *Patterns of Urban Life* by R. Pahl which has been used as a basis for this summary (iii) section 9, page 73, which in effect includes the social comment on the sociological concepts.

1 *Contrasts between urban and traditional rural society*

(a) **differentiation and stratification.** The idea is generally held that whereas traditional rural society was simple, static and clearly stratified with a clear division between the gentry/squirearchy and the working labourers, the demands of a more complex industrial urban society produced intermediate or middle classes and a complex population, mobile and not so clearly stratified. Is this distinction between urban and rural society still valid?

(b) **mobility and status.** In traditional rural society people were said to 'know their place' and could not change their status which was determined by birth. On the other hand in urban society the old class distinctions were broken down as people were able to achieve status by their skill and effort. What is meant by **ascribed status, achieved status, social mobility**?

(c) **social fields: community and association.** The contrast is often made between the close stable village communities and the looser associations of people living in towns and cities. What do sociologists mean by **community, association, total status, partial status**?

(d) **norms and social conformity.** Robert Blythe in *Akenfield: Portrait of an English village* (Allen Lane, 1969) comments that 'There is far greater control in the village: of parents over children, employers over employees. . . .' R. Morrill says on the other hand that 'Cities can tolerate the aberrant, the non-conformist, the reformer, the critic, that rural society cannot endure.' What is meant by the term **norm** and is there still greater pressure today to conform in the village?

2 *How people live in towns and cities: urban ways of life*

Sociologists attempt to distinguish between the behaviour of groups of people by objective measurable criteria, such as how people spend their time and money, their aspirations, their political and social views, the form of relations between members of the family etc. Factors that influence ways of life include occupation (male and female), wealth, nature of formal education, child rearing practices. What is meant by extended and nuclear families? Of which groups are each characteristic? Which of the above mentioned factors have the most important influence on ways of life?

3 *Where people live in the town: social geography*

R. Park described Chicago of the 1920s as 'a mosaic of little worlds that touch but do not interpenetrate'. The city was seen by **urban ecologists** as a sorting mechanism, allocating people to particular areas as a result of social and economic competition.

How clearly distinct and segregated are **social areas** (areas where inhabitants are socially similar) in British towns and cities today?

What economic and social factors determine the sorting process?

4 *Housing classes and urban leapfrog*

Rex and Moore consider the class struggle over housing as an essential part of the urban **status system**. They distinguish between different types of housing, placing them in a hierarchy of prestige which have clear geographical arrangements in the city. They consider that being a member of one or other housing class is of prime importance in determining a family's life style, interests, associations and position in the urban social structure. **Urban leapfrog** is the shifting of population as people move towards more desirable locations in the city, which involves moving beyond (or 'over') existing suburban development towards the edge of the city.

See *Patterns of Urban Life* by R. Pahl for Rex and Moore's list of housing classes and detailed analysis.

Where are the most desirable residential areas in and around the city, and what changes in prestige have occurred in the past 50–100 years?

5 *Social problem areas*

All social areas have their distinctive problems, but in some cases a combination of factors involve the population in a 'vicious circle of deprivation'. General awareness of the extent of severe poverty, slum housing, ineffective education, neglect, discrimination, ill health and crime is reduced by the segregation of the urban population noted above.

Copius information on this aspect of urban life is available from Shelter.

What is meant by **deprivation, deviance**?

6 *Community and neighbourhood*

The ideal communities are often supposed to have existed in traditional rural societies whose members, inhabiting a limited area for a long period of time, were closely bound together by kinship and common experience. Communities of this sort do nevertheless exist in towns and cities, particularly in working-class areas where the extended family system prevails.

'Community' implies that its members have more in common than mere proximity. The term community as used by most people is what sociologists describe as a **locality-based social network** and consists essentially of people who share common interests and attitudes.

The **neighbourhood principle** has been one of the cornerstones of new town planning ideology; emphasis has been placed on the importance of creating community spirit and harmonious social relationships, particularly between members of different social classes.

What is the difference between **community** and **neighbourhood**?

Examine the distinction that has been made between the community-orientated values of working-class society and status-orientated values of middle-class society.

7 *Controlling the urban social system*

There are two major aspects of social control: (a) the informal processes of socialization by which everyone from childhood learns certain basic rules of social behaviour (b) formal institutionalized agencies of law and government.

What is meant by **institution**?

How are our lives controlled by local government?

4 New towns

1 Background: study topics

(a) nineteenth-century enlightened paternalists: Robert Owen and New Lanark; Titus Salt and Saltaire; George Cadbury and Bournville; Joseph Rowntree and New Earswick

(b) visionaries: the Garden City Movement; the work of Ebenezer Howard; Letchworth Garden City; Welwyn Garden City

(c) progress in Housing and Town Planning Legislation: 1842 Bye-law public health and housing; 1868 Torrens Act; 1909 Housing, Town Planning etc. Act; 1932 Town and Country Planning Act

(d) developments in the years immediately prior to the New Towns Act: the Barlow Report; 1944 Town and Country Planning Act; Abercrombie's Greater London Plan

(e) the New Towns Act of 1946 which made provision for

the creation of Development Corporations to acquire designated sites and build new towns

(f) the first New Town at Stevenage 1946

(g) Town Development Act (1952) which enabled the accommodation of overspill population from large cities (**exporting authorities**) in Expanding Towns (**receiving districts**).

2 New towns: where and why

New towns may be put into three categories, according to their main purpose:

(a) to accommodate overspill from congested cities

(b) to provide improved accommodation and suitable sites for new industrial developments in declining areas

(c) to act as growth points of new industrial development as an instrument of regional policy.

Using summarizing tables and maps (see *Geography of Settlements*, F. S. Hudson and *New Towns Come of Age*) construct annotated sketch maps showing the distribution of New Towns and Expanding Towns, the main purposes for which they were built and the main sources of their population.

3 New Towns: aims and guiding principles

Information on New and Expanding Towns is readily available (see section 14, page 105). The following extract from the ministerial preamble to the Skelmersdale New Town planning proposals (1964) is typical of such statements. This and others should be examined in relation to what has in fact been achieved.

(a) Town planning in the 1960s must have regard to a high level of car ownership and car usage and the communications system must be designed accordingly. One advantage of a town the size of Skelmersdale is that it is possible to plan for full use of the motor car. A new town provides the most ideal conditions in this respect and there can be no excuse for any plan which does not face up to these realities. There must be maximum separation of pedestrians and vehicles throughout the town with a separate footpath system to enable people to walk to the centre, schools or other parts of the town without having to come into contact with main traffic.

(b) The motor car can lead to the destruction of the town as a place of urban settlement, but when it is properly controlled the pedestrian can enjoy his freedom and, in turn, his freedom can be exploited to the maximum extent when distances within the town are short and walking becomes more acceptable. This in turn leads to the concept of the town as a compact urban centre with surrounding recreation areas. A large proportion of the population can be within easy walking distances of the central area and all of them can have access on foot to the areas of open space. The development of these principles leads to the abandonment of the neighbourhood system and the central area can thereby be strengthened and contain all the principal shopping, commercial and public buildings. Local shopping needs can be met by corner shops distributed throughout the housing areas, local centres being provided only where residential areas are more than about ten minutes' walking distance from the main centre.

(c) Industrial sites should be located in more than one area to achieve a dispersal of employment centres, thus easing the traffic problem. Some factories of selected types should also be sited in close relationship to the residential areas to ensure more variety in the town and to be convenient, particularly for female labour.

(d) The town should have a coherent structure which can be easily appreciated by those who live in and visit the town, and a proper urban character should be achieved with the correct relationship between buildings and landscape.

(e) The town should fit in to the larger regional landscape but the limits of building development should be clearly defined to achieve the maximum contrast between town and countryside rather than to encourage a suburban sprawl on the periphery of the town. This is important not only in achieving a distinction between town and country, but in the formation of a strong green belt, an important element in south-west Lancashire.

(f) There should be a good balance of population in relation to age groups, family structure and employment to provide a sound basis for the development of the town particularly at the second generation period.

(g) The programming of the development of the town must be arranged to ensure that, at each main stage of the initial building period, a part of the town will be established that will provide a full range of facilities for the population.

4 The evolution of New Town styles, Expanding Towns and New Cities

Designs for the earliest New Towns evolved from the Garden Cities and subsequent innovative concepts produced between the wars not only in this country but in America, for example by *Radburn* (separation of local access from main arterial roads; the cul-de-sac for privacy, safety and quietness; *Olmstead* ('strip parks', green spaces; footpaths between houses instead of roads) and *Perry* (the Neighbourhood Unit concept). The first New Towns incorporated these ideas in various forms and were generally characterized by extensive open spaces and low density housing. They were arranged in 'Neighbourhoods', each with its own small shopping centre, primary schools, community hall, public houses, etc. Functional zones, e.g. industrial estates, were clearly defined. These New Towns have been described as the **Mark 1** New Towns.

Mark 2 New Towns aimed at higher densities to provide a more compact, less sprawling urban form. Traffic segregation was maintained both in shopping precincts and residential areas. Skelmersdale exemplifies a Mark 2 New Town, though Cumbernauld is the best publicized of this type (see the photographs and maps on pages 54, 55 and 70). In Cumbernauld all housing is within walking distance of the town centre to which it is possible to move from any part of the town without crossing a road. Traffic-pedestrian segregation is achieved by multi-level arrangements. Subsequent developments are less easily classifiable on account of the variety of designs which have been adopted, though in all cases, they may be seen as variants or amalgams of these two basic styles. Increasingly, however, there has been a trend (a) towards Expanding Towns involving the welding of New Town development on to existing towns in order to take advantage of existing urban provision, and (b) towards the linking and integration of a number of smaller settlements, e.g. Telford, which incorporates the existing towns of Wellington, Dawley and Oakengates. Official policy increasingly favours Expanding Towns (governed by existing local authorities) over New Towns (governed by a special authority) on grounds of cost which is usually greater in the case of 'greenfield' New Town developments than in areas where the urban infrastructure is already largely established.

New Cities One of the arguments favouring Expanding Towns is that advantage may be taken of larger agglomerations of population in order that a full range of urban services may be provided more efficiently. Some services can only be provided economically if they are used by a large population (these have been described as 'high order functions' which have high population thresholds). It is not surprising, therefore, that the idea of new cities has been put forward, making good sense at least in terms of the efficient provision of urban services. The South-East Study published by the Ministry of Housing and Local Government includes plans for Solent City (a planned amalgamation of Southampton and Portsmouth) and Milton Keynes. Telford, mentioned above, will eventually include approximately 250 000 people in a relatively dispersed urban unit of city proportions. Other plans have been put forward for Humberside, for Preston-Leyland-Chorley, and for Livingston (Scotland) to be developed into a linear city of 250 000.

The linking of existing towns, and new developments along directed lines of growth have much to commend them in the regional context. They provide a more rational provision of services for the population and act as growth points which will collect to them most of the expansion which will occur within the regions where they are situated. But the idea, of course, is a long way removed from the small rural/urban idylls of the early New Town protagonists and may create problems which planners have worked for years to combat.

5 The future: what can we expect?

In our own lifetimes we are likely to see considerable changes in our environment. With the increasing realization that successful planning depends on accurate prediction of future conditions, the science of Futurology has emerged as a serious enterprise.

What changes may have to be taken into account?

The European Cultural Foundation's Project **Plan Europe 2000** is recommended as a source of valuable ideas and information (see Further Sources of Information, page 105). Some of the following examples of possible changes which may affect Britain by 2000 A.D. are derived from this source:

1 100 per cent increase in road vehicles
2 new high-speed mass ground transportation
3 substantially increased life-spans
4 improvements in contraception, administered on a bulk basis through the public water supply
5 widespread development of mind-influencing drugs designed to eliminate undesirable personal traits
6 new attitudes to economic growth (e.g. is it inevitably a good thing?)
7 new attitudes to the traditional concept of a life-long marital union
8 new attitudes to human relations (e.g. decreasing tolerance of coercive bureaucratic arrangements)
9 new attitudes to the balance of personal and collective spending

Project work might involve consideration of the desirability and implications of these and other possible or likely changes.

New urban forms

In the context of new conditions, many projected forms of urban arrangement, which at the present time seem impractical and undesirable, may become acceptable.

Project work:

1 collect information about future possibilities assessing likely disadvantages and advantages:

(a) in terms of today's attitudes
(b) in terms of possible future attitudes

2 draw up plans of your own ideal city of the future

Descriptions of the types of city listed below may be most conveniently found in *Geography of Settlements* by F. S. Hudson (Macdonald & Evans 1970) and Jackdaw No. 80 *Man and Towns.*

The Dispersed City
The Ring City
The Linear City
Satellite and Radial Cities
The Underground City
The Sea City
The Tower City
Plug-in City

The most revolutionary of these are the Tower City and Plug-in City. The Tower City, consisting of a three-kilometre-high tower, built around a 200-metre-square concrete core, with 850 levels housing 300 000 people and containing factories, shops, schools, churches, etc. is already technically feasible. Plug-in City consists of a high narrow grid or mesh, containing a power source, along the top of which would be a mono-rail system for carrying passengers, and mobile cranes for transporting shops, offices, living units, etc. which could be moved and plugged in anywhere along the grid. This might extend from Liverpool to Dover in an early phase of development with later extensions linking up with Paris, Dijon, Lyons and Marseilles, or running across northern and central Europe.

6 Examining facts and figures

The data included here is designed to assist an understanding of general principles and relationships.

It is suggested (a) that the statistics, which have been deliberately left only partially processed, are translated into concise written summaries and/or appropriate visual form (e.g. graphs, age pyramids or maps) and (b) that the following questions are considered:

(i) what do these figures imply and what general concepts do they illustrate?
(ii) if these are the figures for 'x', what are the figures for 'y'? (Where 'y' is some other place or period in time).

Estimated frequency of visits to types of shops		
	grocers	*8 visits per week*
	greengrocers	*4*
	butchers	*3*
	bakers	*3*
	women's wear	*1 weeks per visit*
	chemists	*1*
	hardware	*2*
	men's wear	*4*
	footwear	*7*
	radio/electrical	*9*
	furniture	*32*

(From 'Supermarket and Society', Sociological Review, Vol 10)

Estimated population requirements of several well-known firms		
	Boots the Chemist	10 000
	Mac Fisheries	25 000
	Baratts	25 000
	Sainsbury (medium self service)	60 000
	Marks and Spencer	50–100 000
	John Lewis (Supermarket)	50 000
	John Lewis (Dept. Store)	100 000

(From 'Field Work in Urban Areas', M. Collins, quoted in *Frontiers in Geographical Teaching,* R. J. Chorley and P. Haggett, Methuen, 1970)

A version of the settlement hierarchy concept applicable to England and Wales

Hierarchy status		*Type of settlement*	*Approximate population examples*
1st	Lowest Order	Hamlet	below 50
2nd	Low Order	Village	50–2000
3rd	Middle Order	Town	2000–100 000
4th	High Order	Large town/small city	100 000–250 000
5th	Higher Order	City	250 000–500 000
6th	Highest Order (National)	Metropolis	500 000–5 000 000
7th	World-wide Order	Super Metropolis or World City	5 000 000 and over

40 In the section entitled 'The urban nucleus and its surrounding field' on page 16, it is suggested that the quality of a town's service provision is closely related to its size. This applies to all services – professional, commercial, recreational, etc.

How well is this idea reflected in the status of football teams? The larger the town, the greater is the potential number of paying spectators and wealthy directors to provide the means of obtaining better facilities, a better manager, better players, etc. To test this out – in a very approximate way – rank the towns whose teams play in the football league by population, making appropriate adjustments in the case of towns/cities which support more than one team, and assess the correlation between population and quality of service as indicated by the division in which their football teams play. Which teams appear to be misplaced? What is the minimum population (i.e. threshold population) required to support the service of a league football team? The information derived may be effectively presented in map form.

First division	*Second division*	*Third division*	*Fourth division*
Leeds Utd	Norwich C	Aston Villa	Brentford
Manchester C	Q P Rangers	Bournemouth	Scunthorpe Utd
Manchester Utd	Millwall	Notts Co	Southport
Derby Co	Birmingham C	Brighton	Grimsby Town
Sheffield Utd	Middlesbrough	Swansea C	Southend Utd
Wolves	Carlisle Utd	Rotherham	Lincoln C
Arsenal	Sunderland	Plymouth A	Workington
Tottenham H	Burnley	Chesterfield	Gillingham
Chelsea	Preston NE	Shrewsbury T	Cambridge Utd
Liverpool	Luton T	Bristol R	Doncaster R
Stoke C	Swindon T	Oldham A	Colchester Utd
West Ham Utd	Blackpool	Blackburn R	Bury
Everton	Sheffield Wed	Wrexham	Northampton
Leicester C	Charlton Ath	Port Vale	Reading
Coventry C	Bristol C	Transmere R	Aldershot
Newcastle Utd	Portsmouth	Bolton	Peterborough Utd
Ipswich T	Oxford Utd	Rochdale	Newport Co
Southampton	Orient	Bradford	Chester
Huddersfield	Fulham	Halifax T	Darlington
Crystal Palace	Cardiff C	Walsall	Exeter C
West Brom	Hull C	York C	Stockport Co
Nottingham F	Watford	Barnsley	Crewe Alex
		Torquay Utd	Barrow
		Mansfield T	Hartlepool

League tables Jan 23rd 1972

	Percentage of employed population in non-agricultural activities	*Percentage of total population classed as urban*
UK	96	85
USA	88	70
Australia	85	80
USSR	65	50
Italy	65	55
Japan	60	55
Mexico	45	45
Brazil	45	35
India	30	20
China	20	15
W. Africa	10	10

(From *Geography of World Affairs*, J. P. Cole, Pelican, 1964)

	Percentage of population classified urban	*Gross Domestic Product per capita ($ US)*
USA	70	2790
New Zealand	64	1617
Australia	82	1533
Italy	48	776
Japan	64	589
UAR	38	116
India	18	76
Indonesia	15	69

(From *Patterns of Cities*, A. J. Rose, 1967)

'Taking only the cities of 100 000 or more inhabitants, one finds that in 1600 their combined population was 1·6 per cent of the estimated population of Europe; in 1700, 1·9 per cent and in 1800, 2.2 per cent. With industrialization, the transformation was striking; by 1801, 10 per cent of the people of England and Wales were living in cities of 100,000 or larger; this proportion doubled in 40 years and doubled again in the next 60.'

In general, the later each country became industrialized, the faster was its urbanization.

(From *Cities*, Scientific American Publication, Penguin, 1967)

42 *Cities of the Ancient World*

	estimates of population reached
Babylon	80 000
Uruk	50 000
Athens	100 000
Rome	200 000
Roman London	30 000

(From *Urban Geography*, J. H. Johnson, Pergamon, 1967)

Regional contrasts in degree of urbanization in 1955

per cent living in towns over 20 000	
Oceania	47
North America	42
Europe	35
USSR	31
South America	26
Central America	21
Asia	13
Africa	9

Rapid increase in millionaire cities

1850	2
1900	11
1940	51
1961	80

(From *Millionaire Cities of Yesterday and Today*, D. Linton)

Population

London population changes (figures in thousands)

	1938	*1951*	*1958*	*Greater London Target*
Inner Ring	5974	5137	4932	4718
Suburban Ring	2366	2684	2670	2399
Green Belt	977	1322	1554	1288
Outer Country	833	1008	1259	1377
Total	10150	10151	10415	9782

(From *The Future of London*, by E. Carter, Penguin, 1962)

On average, people in Britain move house once every ten years. (From *People in Towns*, by J. Madge, BBC Publication, Further Education Series, 1968).

Only 45 per cent of the population of Paris are natives of the city. (J. Beaujeu Garnier and R. Chabot)

In 1851, Population of the City of London – 128 000
1961, Population of the City of London – 5 000
1901, Population of the County of London – 4·5 million
1961, Population of the County of London – 3·2 million

(From *The Future of London*, E. Carter, Penguin, 1962)

Age Structures 1961

Age Group	*London New Towns % of population*	*England and Wales % of population*	*London (Inner County 1951) % of population (approx)*
0–4	12·7	8·0	8
5–9	11·2	7·1	12 (5–14)
10–14	9·1	8·0	
15–19	5·6	6·9	45 (15–44)
20–24	5·3	6·0	
25–29	8·4	6·1	
30–34	10·0	6·4	
35–39	9·5	7·0	
40–44	7·1	6·7	
45–49	5·7	7·0	21 (45–59)
50–54	4·2	7·0	
55–59	3·2	6·5	
60–64	2·5	5·4	13 (60 and over)
65–69	2·0	4·4	
70–74	1·6	3·4	
75–79	1·0	2·4	
80–84	0·6	1·3	
85 and over	0·3	0·6	

(From *The South-East Study*, 1961–81 HMSO)

Number of men per 100 women in urban population

England and Wales	90	Canada	97·8
Sweden	92	USA	94·9
Spain	86·5	Brazil	91·4
Iceland	89	India	116

(From *UN Demographic Yearbook*, 1960)

44 Sprawl and concentration

Populations, areas and densities of British Conurbations (Census Definition)

	Area (sq. miles)	*Population 1951 (thousands)*	*Population 1961 (thousands)*	*Density 1961 per sq. mile*
Greater London	721	8346	8172	11300
S.E. Lancashire	379	2421	2427	6400
W. Midlands	268	2236	2344	8750
W. Yorkshire	484	1692	1703	3500
Merseyside	150	1382	1386	9250
Tyneside	90	835	852	9500
Clydeside	324	1759	1802	5550

(From *Geography of Settlements,* F. S. Hudson, Macdonald and Evans, 1970)

Densities are from areas which include open land, industrial areas and some areas of relatively dispersed population.

Between 1900 and 1960, the proportion of urban land in England and Wales doubled to 11 per cent of the total area; it is predicted that this proportion will rise to 16 per cent by 2000. During the 1930s, 60 000 acres of farmland per year were taken by housing.

Projecting the 1900–1960 population growth curve for Los Angeles produces an estimated population of 80 million in the year 2000. A figure of 66 million has been estimated for the population of Calcutta by the end of the century (U K population 1971 – 56 million).

(From *Geography of Settlements,* F. S. Hudson, Macdonald and Evans, 1970)

Densities per square mile

London Administrative County inner Zones	26 000
Manhattan	53 000
Manhattan (excluding Central Park)	82 000
in parts of Hong Kong and the Casbah of Algiers	500 000

Population densities recommended for Central London in 1951 ranged from 14 500 to 29 000 per square mile.
(From *Urban Geography,* J. B. Garnier and G. Chabot, Longman, 1967)

Land values In 1963 an advertisement designed to encourage office dispersal included information to the effect that it cost £2.25 rental and £0.40 rates for floor space for a waste paper basket in Central London.
One hectare (2·471 acres) of land in the Avenue de Montagne has the same value as 8000 hectares of fertile wheat land in the Soissonais and the same value as a third of the whole department (equivalent to county) of Basse-Alpes.

(From *Urban Geography*, J. B. Garnier and G. Chabot, Longman, 1967)

The most expensive land in the world is in the City of London. Values of £3000 per square metre are not uncommon for good sites.

(From *The Guinness Book of Records*, Guinness Superlatives Ltd)

Urban buildings The largest commercial building in the world by area, is the Merchandise Mart in Chicago, built as a wholesale buying site. Its average daily working population is 20 000.

(From *The Guinness Book of Records*, Guinness Superlatives Ltd)

The New York Empire State Building contains offices of 940 firms altogether employing 25 000. 1000 people are permanently employed on maintenance.

(From *The Guinness Book of Records*, Guinness Superlatives Ltd)

Open space per inhabitant in square metres

Paris	1·5
Lyons	4·0
London	10

Paris provides approx. 1 square metre of recreation ground per 15 children.

(J. B. Garnier)

In 1960, 18 million Parisians consumed	250 million gallons of wine 400 000 tons of milk 1 million tons of potatoes 700 000 tons of green vegetables 700 000 tons of meat

Paris produces 1·2 million tons of domestic refuse per year and has 12000 miles of sewers.
(G. Chabot and J. Beaujeu Garnier)
Municipal authorities in England and Wales provided 700 million gallons of water to urban areas in 1900, and 2500 million gallons in 1967.

A 'normal' figure for individual water consumption in Britain is 10–15 gallons per day (for drinking, cooking, washing and sanitation).

Pollution In 1966, motor vehicles in Britain put out 5 million tons of carbon monoxide.

In Paris, where, in the mid-nineteenth century, urban planning envisaged the presence of 50 000, there are now over a million vehicles and the total rises by 140 000 a year. At pavement level, along the main roads, the air in Paris now contains 92 litres of carbon dioxide per 100 cubic metres; the safe limit is 100 litres per 100 cubic metres. Blood samples taken from individuals who have been stationed for several hours at a traffic junction have shown the danger to health.

(J. B. Garnier)
In 1972, specially-designed air purifiers were set up in Central Paris.

Forshaw and Abercrombie estimated before the war that losses resulting directly from traffic delays were about £70 million annually. This figure is now estimated at £500 million. Between 1909 and 1958, vehicle journey speed in Central London decreased from 11·4 to 10·3 mph and the total journey time spent waiting at intersections increased from 25 to 36 per cent. At 27 principal crossings in Central London, an estimated 18 800 hours were lost daily in 1958.

34 per cent of families in Great Britain owned at least one car in 1964.

Growth in number of motorized vehicles 1910–2010 in Britain

Date	*Vehicles per head*	*Total*
1910	·01	
1930	·05	
1950	·09	
1970	·33	14 million
1990	·52	28 million
2010	·55	40 million (estimated population 74 m)

(From *Traffic in Towns*, C. Buchanan, Penguin, 1964)

Housing and social aspects

Percentage of houses classified as unfit for human habitation in 1954 slum clearance returns

Liverpool	43 per cent
Manchester	33 per cent
Birmingham	16 per cent

Age of accommodation units in England

	Owner-occupied	*Local authority rented*	*Privately rented furnished*	*Privately rented unfurnished*	*Other*
pre 1919	40	7	75	76	52
1919–44	35	29	20	17	14
1945–60	14	52	3	4	20
1960–	10	12	2	3	10

(From *Housing Survey*, 1967)

Amenities lacking out of total housing stock of 15·7 million in 1969

		per cent
W.C.	415 000	2·6
Inside W.C.	2 504 000	15·9
Fixed bath	2 106 000	13·4
Wash basin	3 040 000	19·4

Percentage living at more than 1·5 persons per room in British conurbations

Tyneside	4·3
Merseyside	4·2
Central Clydeside	15·2
W. Yorkshire	3·0
W. Midlands	4·2
S.E. Lancashire	2·5
Greater London	4·6

48 *Redevelopment in the Gorbals*

Example of planned change in land use and population

	1956	*Planning proposals*
Dwellings	7 600	3 600
Population	27 000	10 000
Persons per room	1·9	1·0
Schools	6 on 6 acres	4 on 19 acres
Shops	450 (50 vacant)	60 (in 4 centres)
Pubs	48	9

Two examples of the copious statistics given in British Towns *by C. Moser and W. Scott*

Social composition and political affiliation of extremes out of 157 urban areas with populations exceeding 50 000.

Percentage of occupied and retired males in Social classes 1 and 2		*1951*	*Percentage of votes cast for Labour and other left-wing parties*	*1955 General Election*	
Lowest	Dagenham	7·7	Rhondda	80·9	Highest
	Warrington	7·8	Merthyr	77·2	
	Bootle	8·0	Dagenham	73·9	
	Rhondda	8·4	West Ham	73·8	
	St. Helens	8·5	Barnsley	72·8	
	Median	**15·8**	**Median**	**48·8**	
	Wanstead	36·5	Torquay	25·4	
	Esher	37·4	Coulsdon	25·2	
	Southgate	40·0	Esher	24·8	
	Beckenham	41·2	Worthing	22·9	
Highest	Coulsdon	45·4	Southgate	20·0	Lowest

(From *British Towns*, C. Moser and W. Scott, Oliver and Boyd, 1961)

7

Examining maps

The map is a basic piece of equipment for the geographer and planner. A visit to any town planner's department will indicate the variety of uses to which maps are put in portraying a vast range and quantity of data which would otherwise be unmanageable. The maps shown here are selected to illustrate sections of the text, and questions are posed to direct attention to their salient features.

1 Major urban centres in Britain and their spheres of influence

This map of urban catchment areas gives a better idea than a map showing towns and cities alone, of the extent to which Britain is urbanized in the sense of its population participating in the regional division of labour and the provision of services organized about urban centres. It should be noted that areas not covered by the commuting zones indicated, also contain towns with their own catchment areas including Bedford, whose catchment area lies between those of Peterborough (Pb), Cambridge (Cb), Luton (Lu) and Northampton (No).

Consider other methods of showing urban regions.

Using this as a base map (which may be traced): (a) plot the distribution of dormitory towns, New Towns and resorts (b) draw in the motorway network.

Consider the implications of this map for the re-organization of Local Government. (Refer to recent reports.)

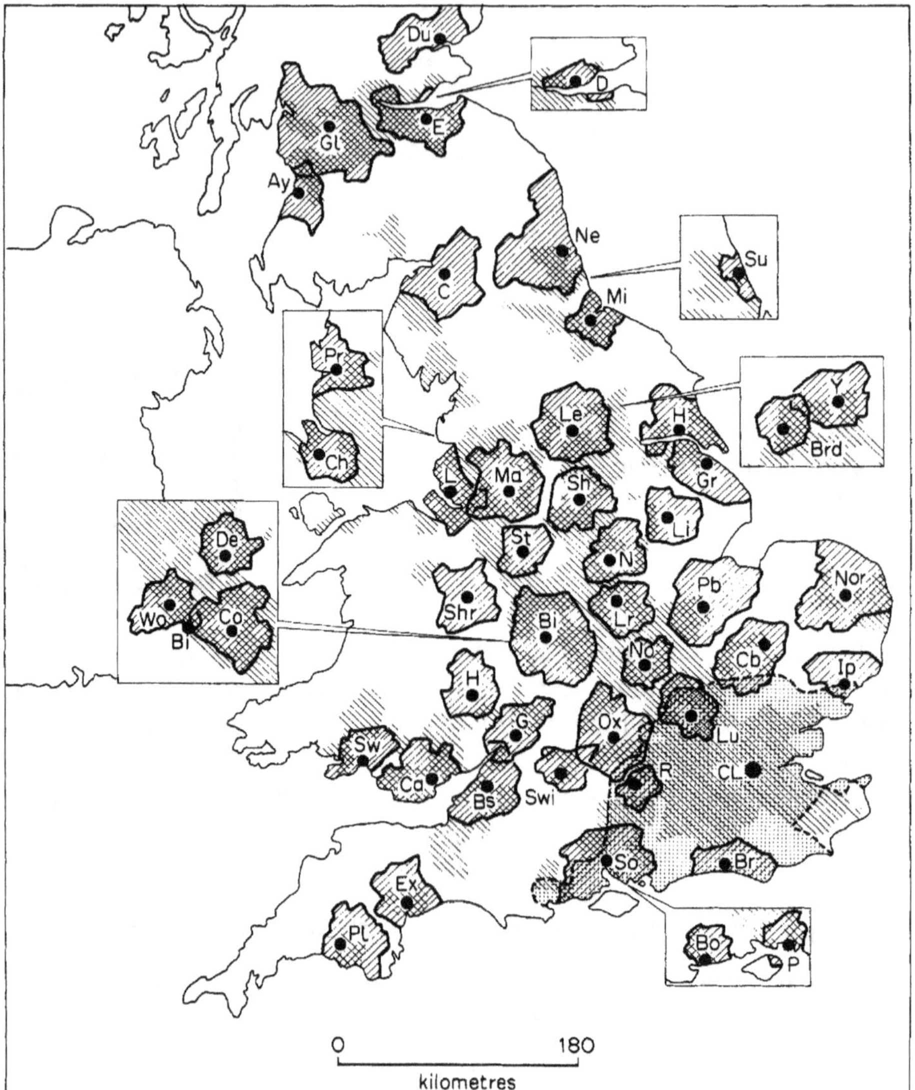

Fig. 16

Urban catchment areas. In 1961 local authorities within the boundaries indicated sent over 100 workers to the town marked. Central London drew 1 250 000 from a wide area. Dependence of many places on a few areas of job concentration is shown by shading. In these, daily out-movement involved over 40 per cent the resident workers.

2 The urban hierarchy in England and Wales

Examine this map in conjunction with the version of the urban hierarchy included in the statistics. How closely do they relate?

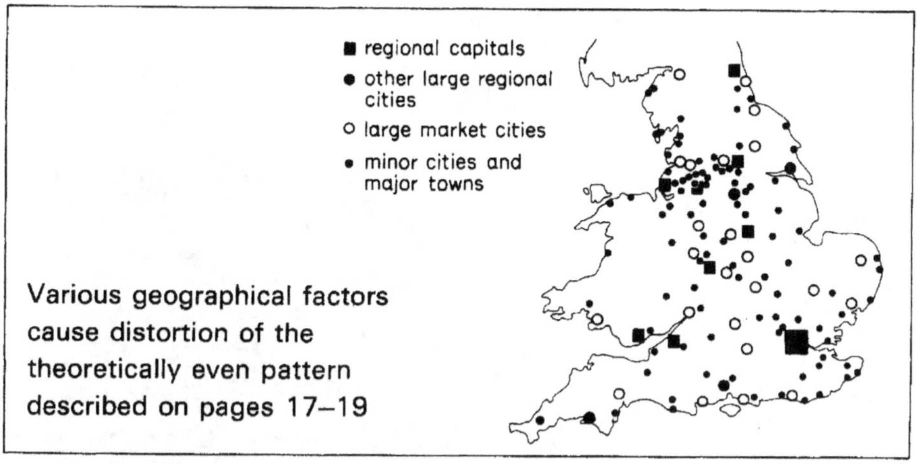

Fig. 17 Urban hierarchy in England and Wales

3 Bedford and Cumbernauld

See questions relating to photographs of Bedford and Cumbernauld on pages 69 and 70. The suggestions below are examples of work that can be done in your own town. Where suitable the questions, which mainly relate to Bedford, should be applied to Cumbernauld.

With reference to 'Urban components and their arrangement' identify the types of land use shown. By use of an 'overlay-trace' draw a land use map, dividing the town into zones (CBD, industrial zone, etc.). Comment on the distribution of elements and their relationships.

Does Bedford show a 'concentric' or 'sector' pattern?

What do you consider might be particularly responsible for Bedford's traffic problems?

What has happened or what will happen to the settlements of Kempston, Goldington, Elstow, Biddenham and Clapham in relation to urban growth?

Reservoir
Lower Farm
Gravel Pit (Dis)
FB
Rests
Water Works
Hoo Farm
Cemetery
Foster's Hill
FB
FB
Ws
138
Sanatorium (Bedford Sch)
THE PARK
The Lodge
Dye Works
Golf Course
FP
Cricket Ground
West Lodge
Windmill Hill
Schs
139
CH
Biddenham Turn
GP
H M Prison
W
FP
BEDFORD
Queen's Park
Recn Gd
Honeyhill
RIVER OUSE
Cox's Pits (Allot Gdns)
Works
FP
F Ps
Hospls
Austin Canons
South End
Barracks
Kempston Grange
KEMPSTON
Hill Grounds
FP
F B
Schools
Manor House
Works
New Town
Springfield House
F B
Bunyan's Cottage
Up End
108

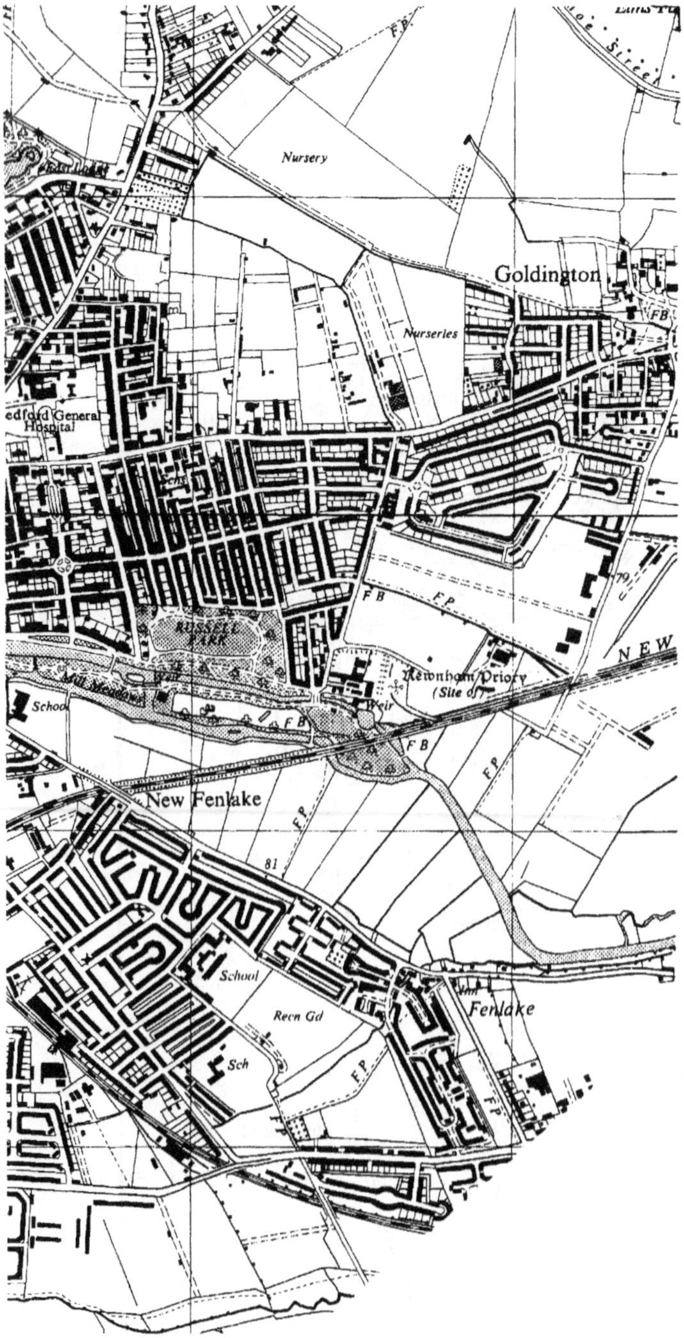

Fig. 18
Plan of Bedford
(1 250 000)

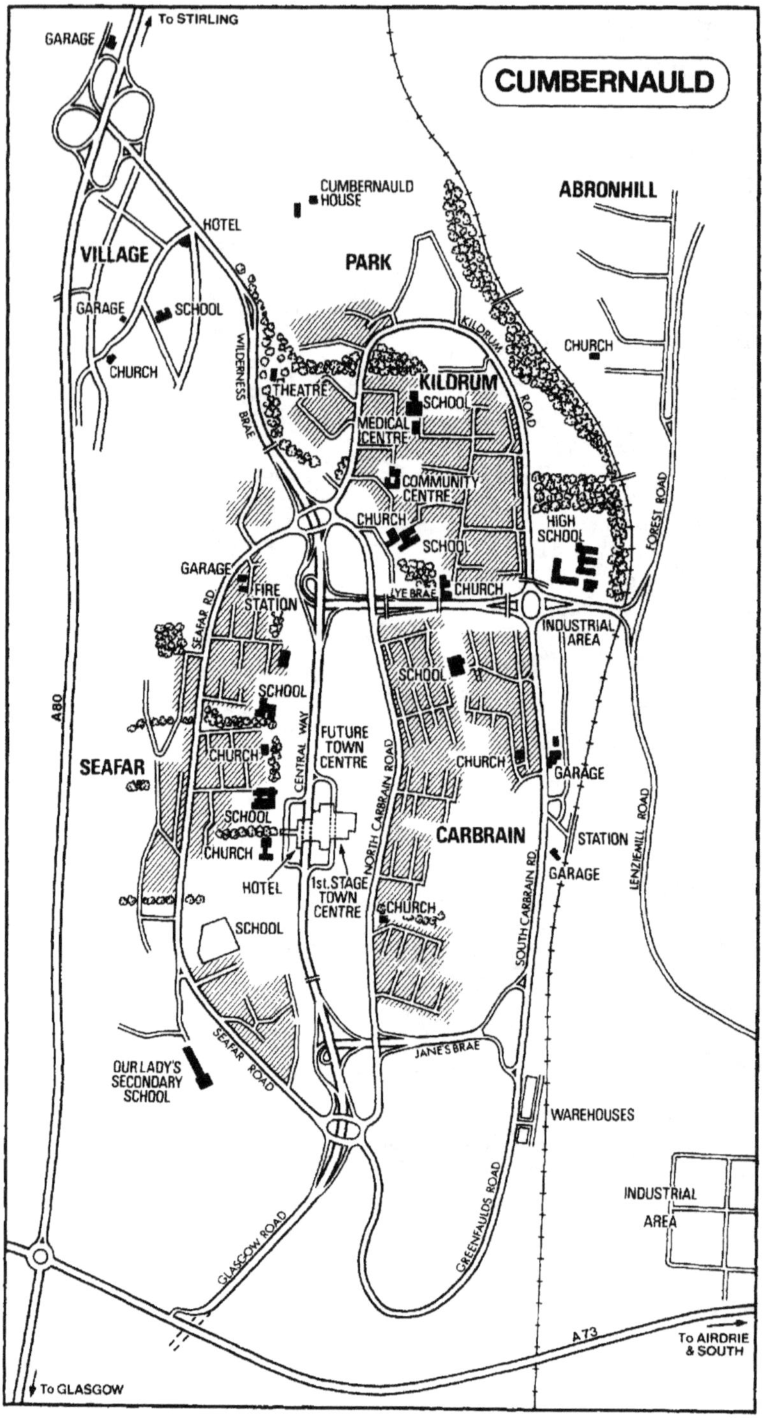

Fig. 19 Plan of Cumbernauld (1 250 000)

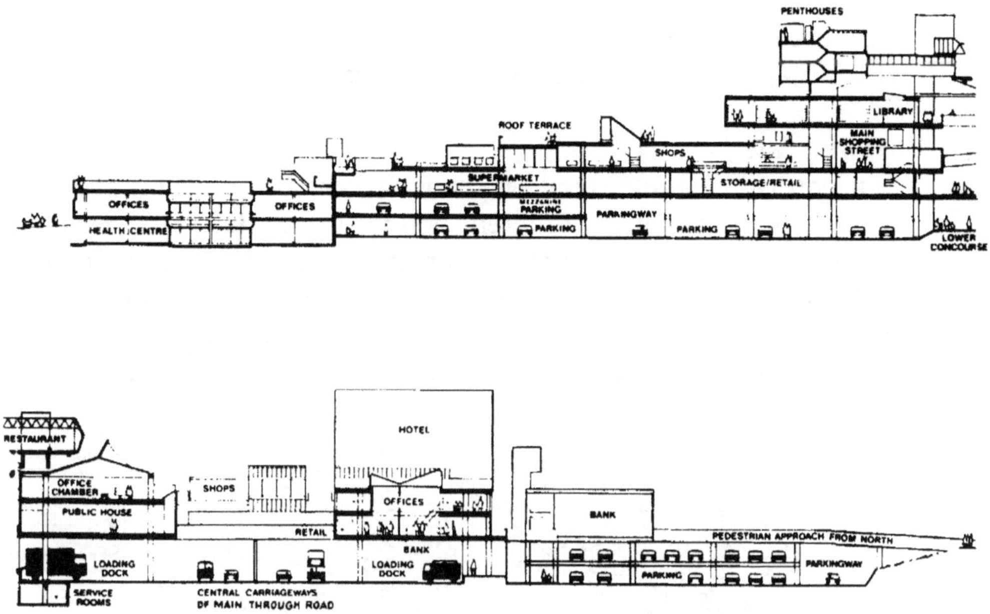

Fig. 20 Section across the multi-level town centre of Cumbernauld

As well as illustrating aspects of urban study referred to in the text, these maps constitute four further examples of types of maps which may be collected and/or constructed for any town or city.

Urban sprawl

Four particular points should be considered: (a) the extent and rate of expansion during different periods (b) the fact that much of London's expansion is not shown on the limited area of this map (c) the position of Wembley (see the photographs on pages 65 and 66) (d) the problems associated with the massive extent of urban sprawl in London especially the preservation of a Green Belt.

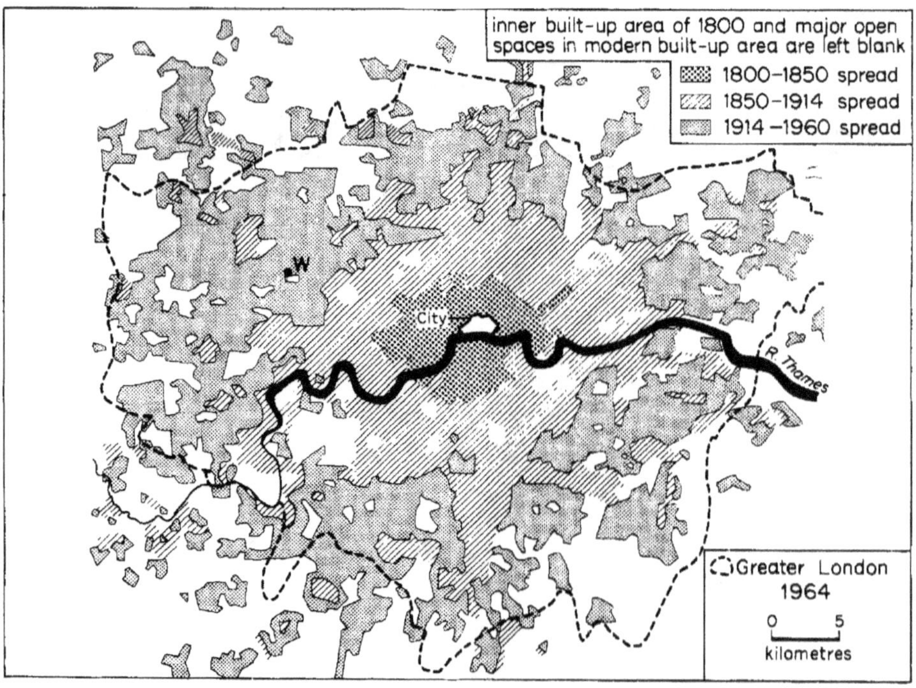

Fig. 21 The growth of London since 1880 (W = Wembley)

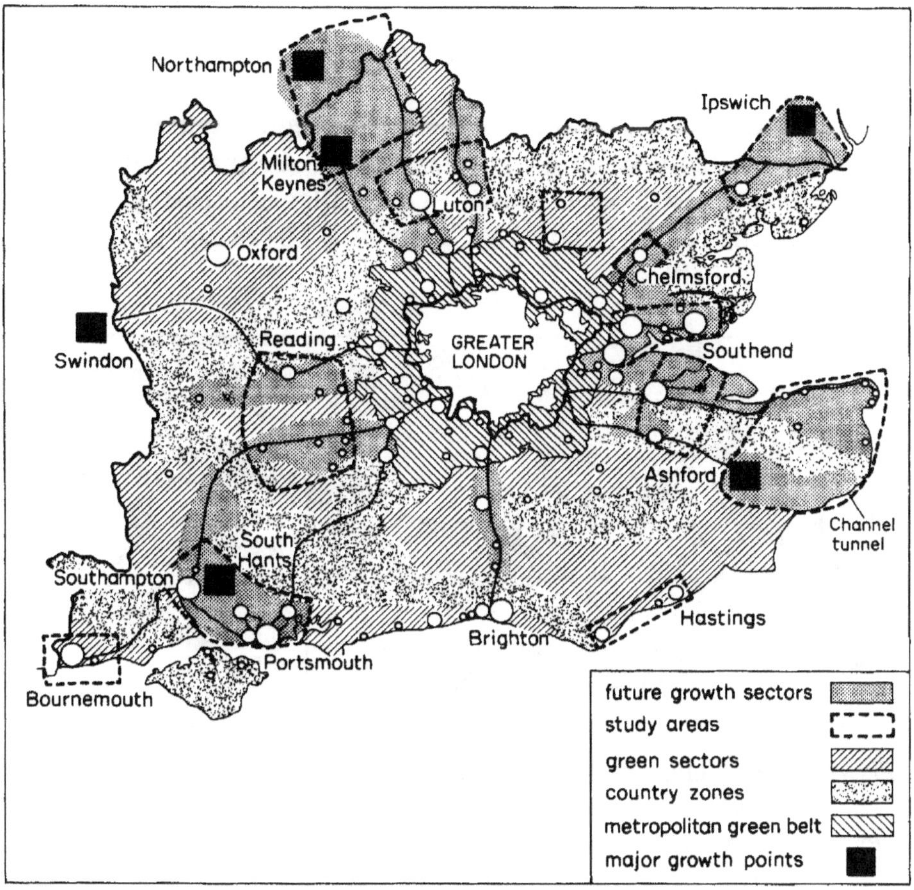

Fig. 22 South-East planning region strategy

Planning strategy

For further details about planning strategy, which aims to direct growth along clearly-defined zones and to preserve areas for particular uses, references should be made to *The South-East Study*, HMSO and *World Cities*, by P. Hall.

58 *Variations in social class*

Examine: (i) the relationships between social class and industrial areas and

(ii) the major areas of distinctive land use in Central London. What generalizations may be made about the distribution shown? For similar information on the rest of London, see *Patterns of Urban Life* by R. Pahl.

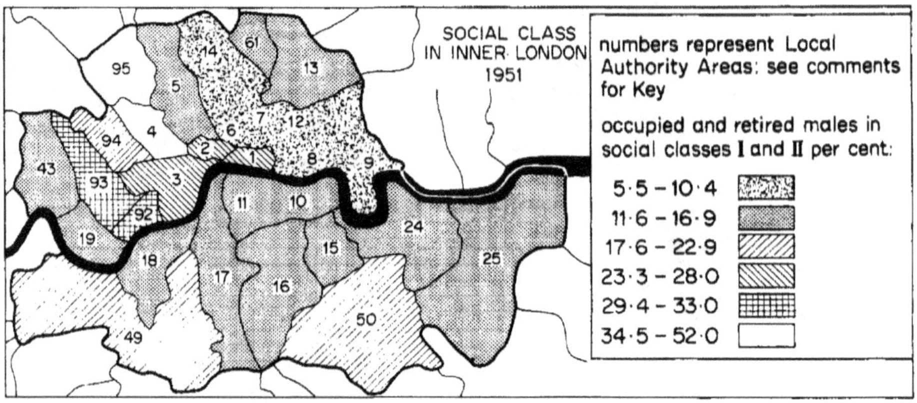

Fig. 23 Social class in Inner London, 1951

Key:

1 City of London	11 Southwark	25 Woolwich
2 Holborn	12 Bethnal Green	43 Hammersmith
3 Westminster	13 Hackney	49 Wandsworth
4 St. Marylebone	14 Islington	50 Lewisham
5 St. Pancras	15 Deptford	61 Stoke Newington
6 Finsbury	16 Camberwell	92 Chelsea
7 Shoreditch	17 Lambeth	93 Kensington
8 Stepney	18 Battersea	94 Paddington
9 Poplar	19 Fulham	95 Hampstead
10 Bermondsey	24 Greenwich	

Population changes across London

With special reference to Understanding the urban environment: urban population, page 26, analyse the factors responsible for the variations shown. Generalize about these variations and consider likely changes in the future.

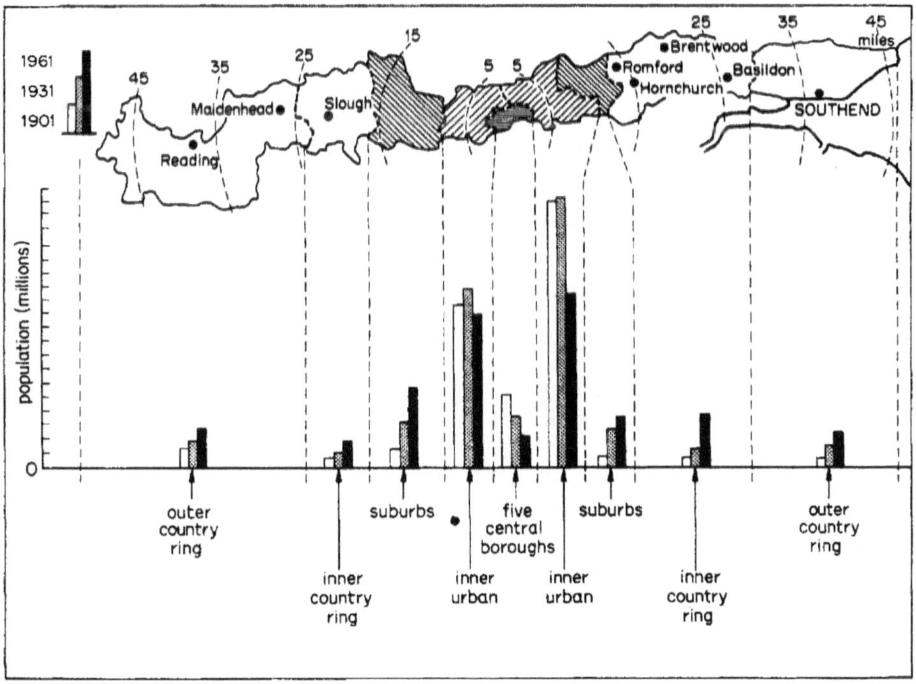

Fig. 24 Population changes across the Metropolitan region

5 Central Bedford

At this scale, 1 : 2 500, street widths and building dimensions are represented sufficiently accurately to allow the degree of precise measurement required in 'field' research. (Larger scale maps are used for special purposes by planners, architects and legal experts.) See also the questions relating to the photographs on page 69 and 70, which require reference to both maps and photographs.

Establish the following: the number of (a) hotels (b) public houses (PH) (c) banks (d) churches;
the areas devoted to (i) industrial functions (ii) educational functions (iii) government and administrative functions (iv) roads and streets.

Calculate the housing density of the area bounded by Midland Road, River Street, Commercial Road, Battison Street.

046
047
048
Police Station
GREYFRIARS
Posts
THURLOW STREET
The Greyfriars (P H)
Church Square
JAMES STREET
Cinema
PRIORY STREET
Car Park
GWYN STREET
Bedford Modern School
Club
MIDLAND ROAD
RIVER STREET
COSTIN STREET
MAITLAND STREET
BATTISON STREET
HORNE LANE
COMMERCIAL ROAD
Cattle Pens
Cattle Market
Ward Bdy

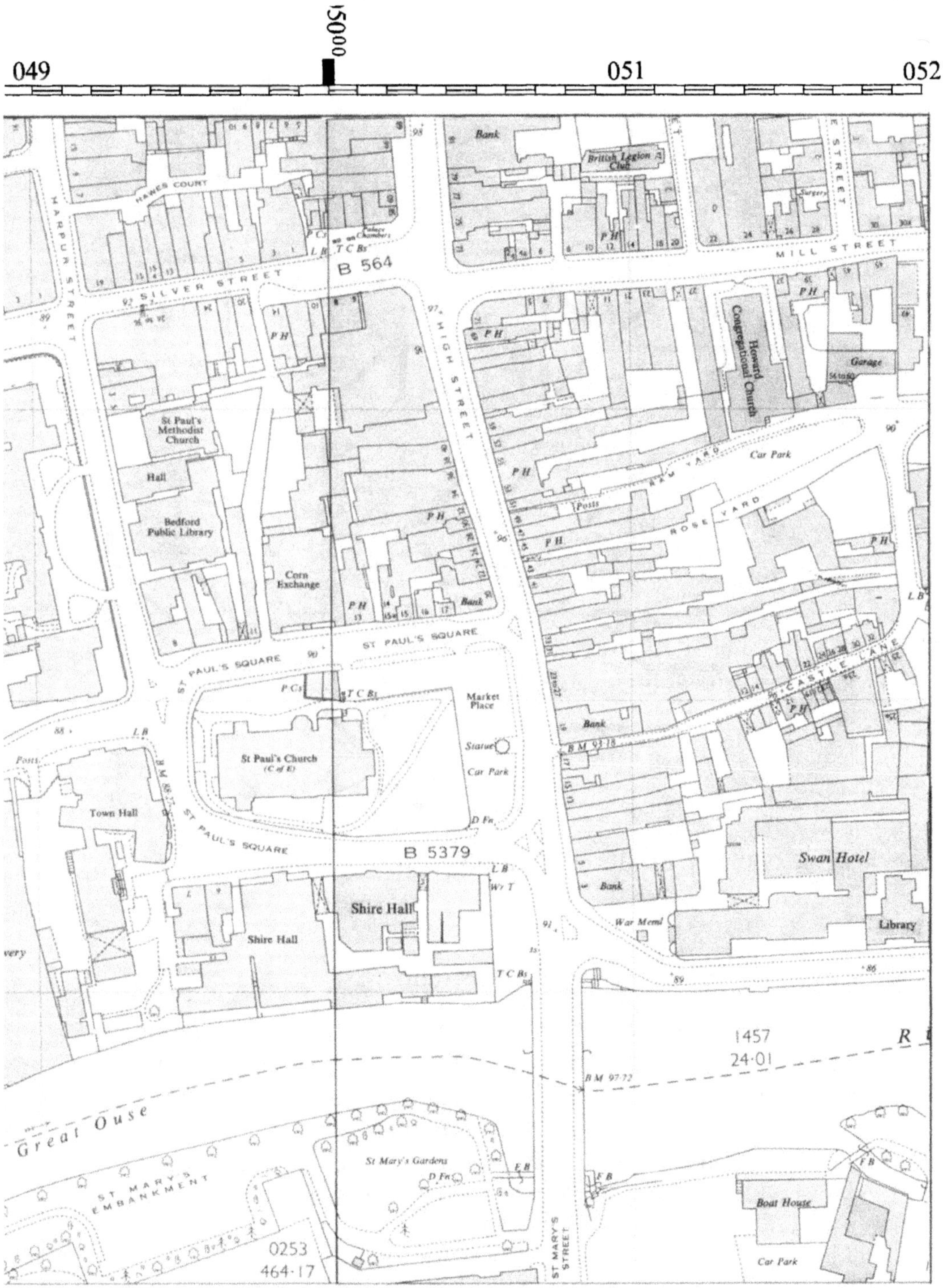

Fig. 25 Plan of Central Bedford (12 500)

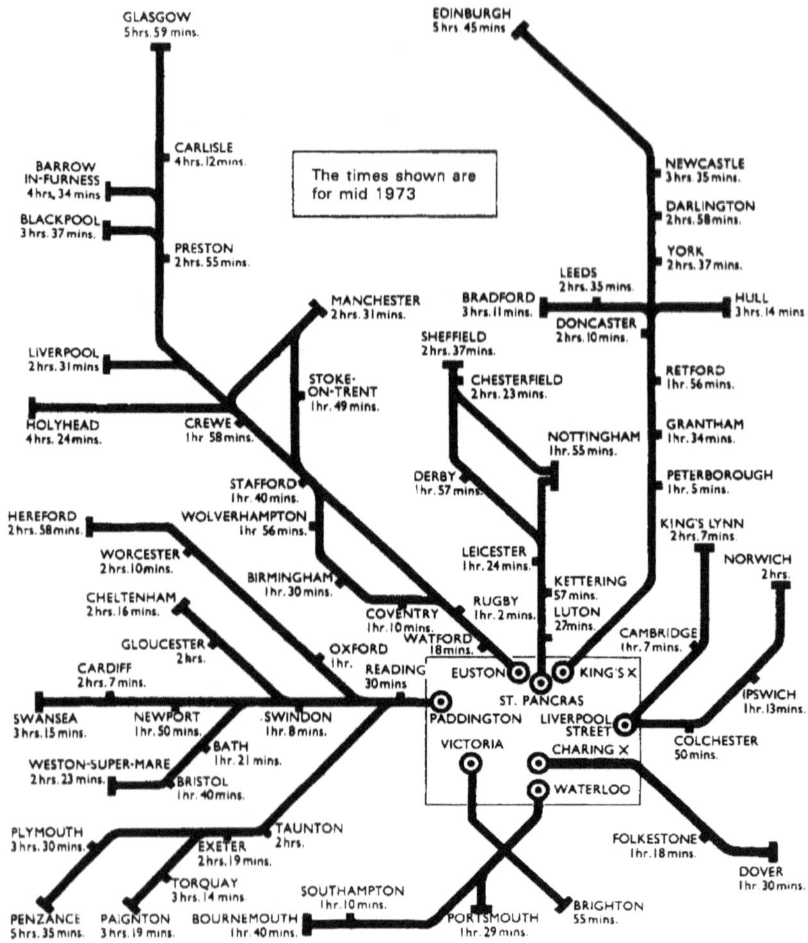

Fig. 26 Suburbs of Metropolitan Britain?

6 The suburbs of Metropolitan Britain

At the top of the urban hierarchy in a country is the primate city' which is economically and culturally dominant, providing a unique range of specialist high-order services, in many cases for the whole country. Britain's primate city is London.

What will be the implications of the reduction of times shown when, with the introduction of advanced passenger trains, average speeds of 150 mph are attained?
Is it reasonable to suggest that Britain's provincial towns and cities are merely London's suburbs?

Examining photographs

8

How meaningful a view is to an observer depends upon his or her appreciation of relationships – between the various visible elements and between the visible elements and the processes they reflect. Any townscape has its economic, demographic, social and geographical as well as its architectural aspects. This applies as much to those parts of the town that are ugly, unproductive and devoid of people (e.g. a slum demolition site) as, for example, to a bustling central shopping area with imposing buildings. With the aim of composing a set of urban portraits, the observer should have an eye for the age, appearance, etc., of both the buildings and the people.

Make a collection of writings on the town by well-known authors (e.g. Dickens, Bennett, Orwell, Sillitoe)

Make a portfolio of sketches, paintings and/or photographs of different aspects or parts of the town, attempting to convey the character and quality of various scenes. The sketches could be literary.

1 Urban field simulation: iron filings and magnetic fields

An analogy is drawn between the urban centre with its surrounding sphere of influence (sometimes referred to as zone of attraction), and a magnetic pole with its surrounding gravitational field. Fig. 27 shows the effect of a magnet placed beneath an even surface with an even-random spread of iron filings representing a non-urban environment of independent non-interacting units. The gravitational field is indicated by a re-arrangement of the filings into a centred pattern; the filings tend to gravitate towards the centre as the surface is vibrated. Fig. 28 shows the effect of several magnets producing the hexagonal pattern of a theoretical environment of urban regions referred to in the section on page 18.

Examine, in the light of this model, the terms **nucleation, packing theory, friction of distance, migration.**

64 Set up the same model introducing variations to simulate roads (wires attached to the magnets), towns of different size, relief, surfaces of different type, etc., analysing the effectiveness and limitations of the simulation.

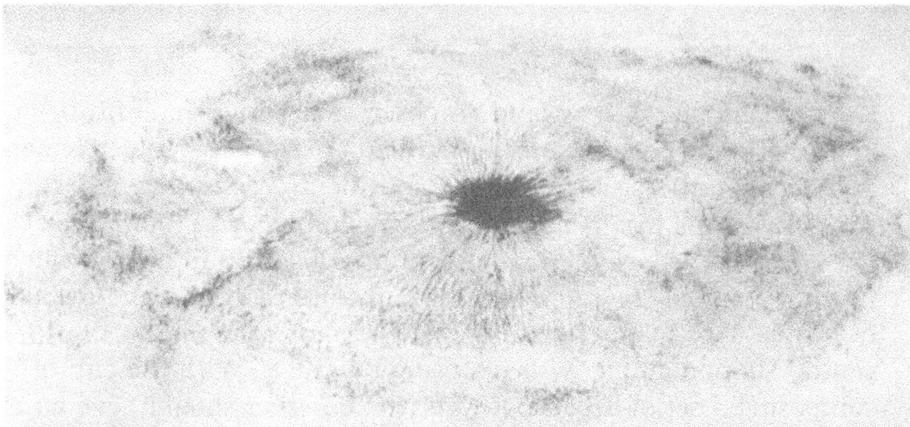

Fig. 27 A surface spread with iron filings: a magnet is beneath

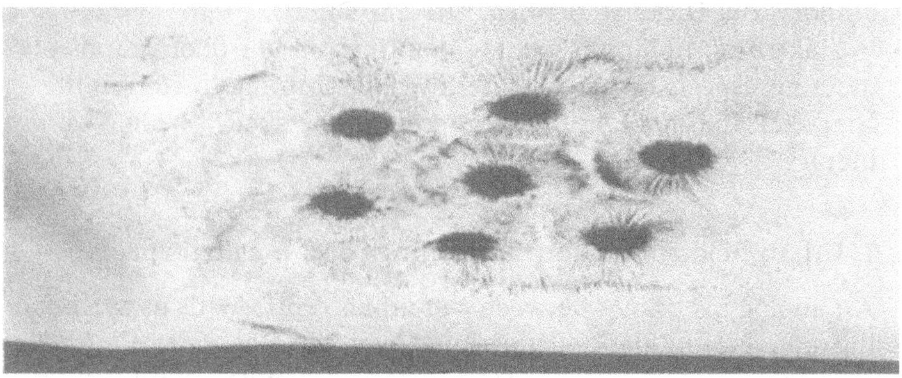

Fig. 28 A surface spread with iron filings: several magnets are beneath

Fig. 29 Aerial view of Wembley in 1921

2 Urban sprawl and urban recreation

Fig. 29 shows an aerial view of Wembley in 1921, less than two years before the first Cup Final was played there. Fig. 30, also an aerial view, was taken during the 1955 Cup Final.

Draw a generalized sketch to show the variety of land use functions on Fig. 30.

Although the main point of this pair of photographs lies in the

Fig. 30 Aerial view of Wembley in 1955

differences between them, what elements of the 1921 scene remain in 1955?

What information does Fig. 30 provide about the processes of urban sprawl?

3 Late nineteenth-century street scene

To preserve old buildings is often to preserve hollow and romanticized images of the past. Fig. 31 shows buildings now vanished from the urban scene, through legislation which has established progressively higher standards of housing and social welfare.

Examine the major advances in housing and social legislation with special reference to the buildings which remain from earlier periods in your town.

What are the current regulations governing building standards?

Do we really derive anything valuable from preserving our 'cultural heritage'?

Fig. 31 Street in Newcastle-upon-Tyne, about 1880

Figs 32 and 33, to be examined in conjunction with Figs 18, 19 and 20, have been selected as representative of two major types of British town. Fig. 32 shows Bedford, a medium-sized regional centre whose form is the product of gradual evolution over a long period of time. Cumbernauld is a New Town fresh from the drawing board. Fig. 33 is an aerial view of Cumbernauld during construction. While, for comparative purposes, the difference in scale of the two photographs must be taken into account, both towns possess the range of functions listed below. Most of these can be identified on both photographs.

Fig. 32 Aerial view of Bedford

Locate on both photographs, buildings associated with residential, industrial, commercial, educational, social and administrative functions.

N.B. (a) What large building has been constructed in Bedford since Fig. 18 was drawn up? (b) Bedford Corn Exchange is now a concert hall, convertible for wrestling (c) the Cumbernauld Town Centre appears insignificantly and only partly constructed in the centre background.

Examine and comment on the contrast in lay-out of roads and streets.

Fig. 33 Aerial view of Cumbernauld

 As in most towns, traffic flow and car-parking are major problems. Identify three different forms of car-parking provision in Bedford.

What problems other than traffic problems may be inferred from examination of the Bedford photograph?

In what way do you consider that 'problems' which the planner may identify in Bedford, constitute part of the character of the town which many Bedfordians would be reluctant to lose?

5 Reality and vision

Fig. 34. Builders of nineteenth-century working-class housing, condemned as indifferent exploiters of the urban poor, were in this case conforming to new legislation which laid down higher standards of construction (compare with Fig. 31). How much of this type of housing remains in Britain today?

Fig. 35. The Central Business District of New York is the classic

Fig. 34 'Regimented monotony' – nineteenth-century terrace housing in Burnley

Fig. 35 New York CBD and Stuyvesant Town (foreground)

example of the skyscraper response to astronomical inner city land values.

Draw a generalized profile (top right to bottom left) to indicate the land value/land use intensity gradient.

Less frequently photographed are the rectangular blocks of Stuyvesant Town, a slum replacement project built partly with state funds. At a density of 393 per acre 'this housing would require eighty additional acres to provide the park and playground space now regarded as desirable'. (L. Mumford)

Fig. 36. Though not quite what Ebenezer Howard had in mind, this may be considered as one variant on the 'back to nature' theme. Constructed to a height of 3 000 metres it is aesthetically distinctive and offers considerable recreational advantages for the rock climber and the athlete who wish to train at altitude. The reader is invited to find a title for it and design an alternative vision of the future.

Fig. 36 Stereoscan 'micrograph' of a lead tin telluride crystal, magnification 60 ×

9

Examining views: quotations on aspects of urban life

Country life versus city air

'God made the country, man made the town.' (Varro)

'Cities are the final pit of the human spirit.' (Rousseau)

'City air makes one free.' (Old German Adage)

'The English are town birds through and through. . . . yet they don't know how to build a city, how to think of one or how to live in one.' (D. H. Lawrence)

'I would rather enough cubic feet of living space and an efficient milkman than three acres of land and a cow.' (R. Frankenberg)

'The townsman envies the villager his certainties . . . one day he will find a cottage on the green and real values.' (R. Blythe)

'Cities are places of maximum human interaction . . . where life is lived at its fullest, where the stresses and challenges of the environment may be great but where the creative response is at its highest.' (*European Studies*, Vol 10)

Integration and segregation

'We want to revive that social structure which existed in the old English village where the rich lived next door to the not-so-rich and everyone knew everybody.'
(Chairman, Stevenage New Town Corporation, 1947)

'The mere shortening of the physical distance between different social groups can hardly bring them together unless at the same time the social distance between them is also reduced.' (R. Glass)

 'It was common to live in the square; it was less common to live in the Breach . . . it was most common, most degraded of all, to live in Parkins Row.' (D. H. Lawrence)

'Until now it has been argued that overcrowding would be reduced and integration achieved by thinning out the number of immigrants in twilight areas in the centre. . . . At Bedford Mr C. S. Osborn told a select committee that housing officers had tried to house immigrants in places not now regarded as immigrant areas, but that very often they had refused.' (*The Times*)

Community versus privacy

'In all the city, there is no man who is so near yet so far from me.' (Martial)

'It is a commonplace of modern sociology to refer to the impersonal rationalised and segmentary relationships between individuals in urban societies.' (T. P. Bottomore)

'It is the small face-to-face groups that matter. Small groups are the bricks from which communities are built.' (J. Madge)

'It is the anonymous and impersonal atmosphere of the large town which makes it so ideal an environment for deviant sub-cultures to flourish.' (J. Madge)

'It is never easy to discover exactly what dream is in the minds of those who talk so fervently about "sense of community".' (W. J. H. Sprott)

'The crushing limiting power which the village exerts on families which never escaped.' (R. Blythe)

Neighbours . . . and friends?

'Five thousand (people) would seem to me to be an upper limit (for the size of a neighbourhood). Even if no further advantages to face-to-face association and friendly intercourse and political cohesion followed from neighbourhood planning, one could justify it on economic grounds alone. In terms of educating the young and of making the institutions of democracy work, the arguments are entirely in favour of mixed communities.' (L. Mumford)

'The siting of houses might provide opportunities for exacerbating antagonism just as much as providing opportunities for friendship.' (L. Kuper)

'In architectural terms, it is important that buildings should promote identity; that is the individuality of each citizen. They should also promote social involvement, the opportunity for contact at every level.' (T. Crosby)

'It does not seem possible to assert that the spatial position of the homes contributes one way or another to the formation of friendships.' (G. D. Mitchell)

High density versus low density

'These crowded cities have done their work. They were the best that a society largely based on selfishness and rapacity could construct.' (E. Howard)

'London is not swinging; it is suffocating. Any plan for the future must give London more breathing space or the heart will fail.' (P. Hall)

'If you go into any common outskirts of the town now and see the advancing army of bricks and mortar laying waste the country fields and shutting out the air, you cannot fail to be struck by the consideration that if large buildings had been erected for the working people instead of the absurd and expensive walnut shells in which they live, London would be about one third of its present size.' (C. Dickens)

'No good can come of the emotional assumption that dense city populations are *per se* undesirable. In my view they are an asset. The task is to promote the city life of city people, housed let us hope in concentrations both dense enough and diverse enough to offer them a decent chance at developing city life.' (J. Jacobs)

76 Housing – who cares?

'The most squalid housing conditions together with poverty, ill health, inadequate services and ineffective education are imposed upon slum families merely because they are weak, disadvantaged and relatively easy to confuse and ignore.' (Shelter)

'There could be up to forty in the class. Perhaps a dozen different races are represented. Maybe a half of the children are from broken homes – from a restlessly changing pattern of divorce, separation, desertion, fostering, widowhood, living with relatives and the rest. Practically all of them will be existing in sub-standard accommodation deprived of that modicum of civilised privacy which centres on regular hot water, a bath and an indoor lavatory. Those who do enjoy these relative luxuries will often live perched monstrously sky-high on a towering precipice of flats. Which is preferable? Is it the multi-let tenement, seven or eight in a room, with possibly no gas or electricity, drains blocked and running water in every room, albeit down the fungoid walls? Or is it the flat, with excrement in the lift, menaces on the stairways, drunks in the play-area, frighteningly low rails on the balconies and a blockage in the refuse shaft? There will be a consortium of medical problems, physical and emotional, together with an endless round of social posers along the lines of truancy, absenteeism and so forth. The low standards of attainment, at least by traditional measures, obtrude almost as an afterthought when the terrifying conspectus of a down-and-out downtown classroom is observed.'

(E. Midwinter)

Do appearances matter?

'People who live in mean despicable surroundings become mean and despicable.' (D. H. Lawrence)

'Around us we see so much that is squalid and depressing. So many of our towns have grown into sprawling urban masses without shape or beauty. Everywhere there are signs of indifference and neglect – garish advertisements, the clutter of traffic signs, broken fences, peeling paintwork, cobwebs of overhead wires, abandoned motor cars, and litter of all sorts.'

(Civic Trust Pamphlet)

. . . our modern cities. They are not beautiful nor are they convenient'. (G. C. Scott)

'They (the public) are clear enough about the ugliness of the world they live in . . . and are quite vocal about the heat and the congestion, the chaos and the monotony of it. . . . But they are hardly aware of the potential value of harmonious surroundings . . . they can have little sense of what a setting can mean in terms of daily delight or as a continuous anchor for their lives or as an extension of the meaningfulness and richness of the world.'
(K. Lynch)

Going faster – but where?

'A city made for speed is made for success.' (L. Corbusier)

'Good communications – the key to economic development.'
(E. Heath)

'We must come to terms with the motor vehicle.' (C. Buchanan)

'Traffic is not important. The first priority is that the motor car is reduced from a necessity to a luxury.' (T. Crosby)

'The freedom with which a person can walk about and look around is a very useful guide to the civilized quality of an urban area.' (C. Buchanan)

Planning – for whom?

'Town planners' leader believes that what the community thinks important deserves attention.' ('Credibility depends on dustbins.')
(*The Times*)

'The planner is often frustrated or annoyed when the residents of his bright new housing block look back nostalgically to the time when life was "really sociable".' (J. Madge)

'The German sociologist E. Scheuch, once asked people what sort of environment they would like to live in. A substantial number wanted a detached house by a lake in its own extensive grounds five minutes from the city centre. The job of the planner is to realize the impracticability of that solution and then do his best to get as near to it as possible.' (*Geographical Magazine*)

'Planning is a prime example of the need for participation, for it affects everyone. People should be able to say what kind of community they want and how it should develop; and should be able to do so in a way that is positive and first-hand. It matters to us all that we should know that we can influence the shape of our community so that the towns and villages in which we live, work, learn, and relax may reflect our best aspirations. This becomes all the more vital where the demands of a complex society occasion massive changes: changes which in some areas may completely alter the character of a town, a neighbourhood or a rural area. The pace, intensity and scale of change will inevitably bring bewilderment and frustration if people affected think it is to be imposed without respect for their views. This leads all too easily to alienation between the authority and people. Some people are bound to be hurt and others will remain dissatisfied even though they are informed of proposals and are able to comment on them. Not everyone's wishes can be met. But the fact that some people may ultimately be hurt only strengthens the need for them to know of proposals early, to understand them and to be involved in shaping them.' (*The Skeffington Report*, 1969)

Tension – is violence inevitable?

✠'Throughout the large cities of the world and in Britain in particular the underlying tensions of city life are breeding ever increasing violence.'

✠'The factors related to urban tension are the presence of a large number of people with severely limited educational and occupational skills, unemployment and underemployment, housing needs, consumer exploitation, special problems associated with youth (their education in general, need for specialized occupational training, recreation, delinquency) and community-police relationships.'

✠'The urban tension of the 1960s seems to result from several trends in American life, including:

the generally improved economic status of American families

a shift from a great need for unskilled workers to a great need for white collar service workers and technicians

increased frustration among the disadvantaged resulting from the increasing discrepancy between expectation of improved status and the hard reality of being unable to achieve this.'

✠'There is the need to develop within especially tension-prone groups a sense of positive identification with the community – the perception that the community does possess positive influences operating to one's benefit and that one's personal needs can find avenues of positive expression within the community. This is in contrast to perceiving the community as threatening, non-supportive and closed in terms of avenues for self-development.'
These four quotations are taken from *Tension in the Cities,*
J. A. Bayton

10

The future: what do we want?

Values survey

The question 'What do we want?' should precede practical planning whose success may be best judged by how far the requirements of the majority are accommodated.

The list of generally desirable objectives given below may be used for a series of surveys. For your study group it may be considered desirable to add to the list, but for surveys of other people who may have less time to spend, a shorter list should be devised by grouping items into major categories.

1 Either by placing items in order of priority or asking respondents

 to state the three most important, establish the concensus opinion of:

(a) members of your study group
(b) male students
(c) female students
(d) the middle aged
(e) the middle class
(f) the working class
(g) town planners.

2 Are there significant differences between the views of different groups?
3 On what basis are planners today making decisions about the future?

Building more new motorways
Improving bus services
Improving surface and underground rail services
Developing new forms or transportation
Improving primary education
Improving secondary education
Providing more recreational opportunities for young people
Making better provision for the handicapped
Giving more help to old people
Building more new homes
Reducing social segregation
Fostering better community spirit
Reducing crime
Reducing population density
Reducing urban sprawl
Improving the appearance of urban areas and preserving buildings of special interest
Controlling air and water pollution
Reducing noise
Improving shopping facilities and building more cut price, discount, stores
Providing more jobs
Segregating housing and industry and providing more open space.

Urban issues: testing political reaction

11

Imagine that representatives of the following groups are on a platform (e.g. in an 'Any Questions' situation) replying to challenging questions from the floor which require that they commit themselves to a definite responsible statement on each of the 22 issues:

1 a member of the Conservative party
2 a working-class Trade Unionist
3 a left-wing 'intellectual' Sociologist
4 a representative of the Confederation of British Industry
5 a member of the Civic Trust.

Complete each unfinished sentence to form brief statements reflecting the essential views of each body. (N.B. The stereotype images which emerge can of course be tested by research.)

1 The most important urban problem of our time . . .
2 The future of urban life in Britain as I see it . . .
3 Overcrowding in our cities . . .
4 The accommodation of increasing numbers in high-rise flats . . .
5 The existence of city slums in the 1970s . . .
6 The increase in crime in cities . . .
7 The urban rat race . . .
8 The segregation of our towns and cities into social class areas . . .
9 The dehumanization of the urban dweller . . .
10 The erosion of the countryside by urban sprawl . . .
11 The ugliness of our towns and cities . . .
12 Spending public money on preserving cities like York and Chester . . .
13 The low level of public participation in local politics . . .
14 The suggestion that town planners have, since the war, created more problems than they have solved . . .

15 Banning cars from urban centres . . .
16 The massive increase in urban motorway building . . .
17 The pollution of our urban environments . . .
18 The lack of social and recreational amenities in our cities . . .
19 The decline of the intimate corner shop in the face of competition from vast supermarkets and hypermarkets . . .
20 The suggestion that slower economic growth and lower levels of personal spending may be necessary to preserve the quality of our urban life . . .
21 The idea of a retreat from urban living in the latter part of the twentieth century . . .
22 The astronomical rise in urban land values . . .

12 Key questions

To answer this set of twenty general questions would give a very wide coverage of most aspects of urban study. While they may be used in various ways, e.g. for lecture or discussion topics, the needs of those who may be writing essays in preparation for A-level General Studies have been borne in mind.

1 'Urbanization is more than the collecting together of population into towns and cities.' What is urbanization, in what circumstances does it occur and how does it affect non-urban areas?
2 When we refer to the Developing Countries we are referring to countries which are developing an urban way of life. Is economic development possible without urbanism?
3 'Cities create civilization.' Discuss.
4 'City life has not only become uncivilized, it has become unliveable.' Discuss.
5 Is there any alternative to urbanism?
6 'The country no longer has a life of its own: rural areas have been reduced to peripheral urban land.' Discuss.

7 'We are all living in the rat race now' (Akenfield village). Can we substitute town/city for rat race?
8 Why is there such a low level of public participation in municipal politics?
9 Is there an ideal or optimum population for a town or city?
10 As urban populations increase, we are faced with either congestion and overcrowding or urban sprawl. Which is preferable and is there any alternative?
11 Was nineteenth-century life as dismal and degrading as we are usually led to believe? What image of urban life is conveyed by Dickens or other nineteenth-century writers?
12 Examine the legacy of the past in the town today. What value has it and what problems has it left us with?
13 Does a study of the past provide us with any useful ideas for planning the future?
14 What changes are likely to come about in our towns and cities in our lifetime?
15 On what basis should urban planners set about the task of planning the future of our towns and cities?
16 Have New Towns been a success?
17 Why do you think transport is often considered the key to an understanding of the urban environment?
18 What are the major problems associated with transport in towns and cities and what long-term solutions are there?
19 What and where are the major social problems of our cities today?
20 'Mental fatigue, boredom, tension, violence, loss of identity, neurosis. . . . Are the psychological problems of urban living being exaggerated or is there a dangerous menace lurking beneath the surface of our everyday lives?

13 The scientific approach: method of enquiry

Scientific enquiry is characterized by accurate observation and measurement of data which is systematically organized to produce objective general statements which give a coherent idea of the complex world.

The social scientist has greater problems than the physical scientist in formulating and testing general statements in as far as the raw data is often more difficult to measure and there are more problematical variables to contend with. This does not mean however that the social scientist is less scientific; rather that he must be doubly careful.

The following outline of procedure and techniques shows how the process of research enquiry should be carried out if acceptable results are to be produced.

1 Context of enquiry

This should consist of a summary of the area of study of which the proposed enquiry forms a part, noting some of the major relationships and concepts which contribute to a general understanding of the subject and the overall objectives of the research and reasons for undertaking it.

2 Statement of hypothesis to be tested or problem to be solved

The aim should be to test an important relationship or set of relationships. The wording should be clear and unambiguous, directing the research to some specific limited objective. (Examples are given in this section, see page 94.)

3 Detailed practical considerations

This stage is likely to accompany the selection of the hypothesis and will involve drawing up a timetabled programme of work, listing sources of information and equipment required, anticipating possible difficulties which may require adopting alternative courses of action, establishing an efficient division of work in the case of a group project, etc.

4 Collecting and recording data

Some data may be obtained from 'secondary' sources (i.e. processed information taken over by the student at second-hand usually from journals, books, etc.). But essentially field research requires deriving information at first hand from 'primary' sources (i.e. raw data derived directly from the environment by physical measurement, questioning members of the public, recording land use, etc.) or 'near-primary' sources (i.e. semi-processed data such as statistical tables, published maps, directories and historical documents).

While the collection of data is often a relatively simple routine task, it has to be undertaken with great care. It is very easy to collect partially correct and misleading information. Some knowledge of sampling techniques and questionnaire design is essential and it is desirable to be equipped in such a way that required information can be noted clearly and economically. One or more of the following may, for example, be needed for recording information in the field: blank outline map or section, field notebook for sketches and recording measurements, suitably-prepared sheets for recording traffic flow or answers to a questionnaire, tape recorder for extended interviews. (For further details see the following sub-section, page 87.)

5 Data processing: storage, classification and presentation

The collected data should be organized in a clear and systematic way by translation into one or more of the following forms:

(a) maps and/or diagrams (e.g. density shading, isopleth, dot and/or flow maps, star or ray diagrams)
(b) transects or 'profiles'
(c) graphs or charts (e.g. linear graphs, bar graphs, histograms, divided circle charts, population pyramids)
(d) numerical summaries (e.g. comparative averages)
(e) concise written statement.

The relationships presented in the form of maps or graphs etc. may appear quite obvious, and for general purposes a written assessment summarizing these relationships will usually be satisfactory. However, conclusions may be established more objectively by, for example, the null-hypothesis test which shows whether differences (between mean figures) are due to chance or not, and the chi square test which measures the degree to which two or more groups of data are related to each other. It should be noted that these tests of significance and correlation do not establish whether the relationships are causal.

The conclusion of the research project will relate to the positive, negative or inconclusive nature of the findings, supplemented by some form of critical or qualifying comment about cause and effect, or – in the case of an inconclusive result – about further enquiry which might be needed to clarify the question.

(N.B. For those interested in pursuing the more refined statistical aspects of the work, reference should be made to the appropriate section under Books and other sources of information.)

7 What next?

What has been achieved? The answer to this question may be that understanding has been deepened. Though an answer of this sort may be the ultimate one for the student as academic researcher it may not be good enough for the student as citizen. Should the research shed light on some important local problem, the researcher may feel prompted to present his findings to those in a position to take some remedial action, e.g. an MP, a council member, a newspaper editor, a representative of a voluntary local organization concerned with the particular problem. Indeed it is at this point of 'community feed-back' that the research effort becomes fully justified to the socially committed.

Before embarking upon a research project it would be profitable and potentially time-saving to fill in the following table, preferably after consultation with those who are professionally concerned with such work (e.g. members of a Local Planning Department).

Information required	*Sources*	*Recording method*	*Method of presentation*
Population densities			
Population migration			
Population trends			
Age and sex structure of a population			
Distribution of racial groups			
Employment structure			
Housing densities			
Age of housing			
Condition of housing			
Land values			
Land use			
Traffic flow densities			
Traffic speed			
Traffic accident distribution			
Journey to work patterns			
Incidence of crime and other social problems			
Noise and pollution levels			
Newspaper circulation area			
Department store catchment area			
Local shop catchment area			

Examples of field research technique: the questionnaire and the land use map

Whatever methods of data collection are used, they should always be designed to collect accurate specific information. Unless minute care is taken, even such apparently simple tasks as counting vehicles or pedestrians and measuring street widths can produce inaccurate and misleading information. For full consideration of a wide range of problems and techniques, it is worth making special mention of two books: *Knowhow* by R. Hare, Intertext, and *Field Work in Urban Geography* by K. Briggs, Oliver and Boyd.

88 *Questionnaire*

The selection of questions below is taken from two actual surveys by (a) Strathclyde University, researching into life in Cumbernauld, and (b) the Urban District Council of Sandy, Bedfordshire, as part of the planning stage of a town development scheme.

In both these cases information was being collected for the benefit of those concerned, who could see obvious advantages in giving full and accurate answers. The private, individual surveyor may not get the same co-operation from people reluctant to have their time wasted, and be pestered for information they may not in any event particularly wish to divulge. In other words, in addition to technical problems, there are problems of dealing with people. These may relate to the general attitude towards the bearer of the questionnaire, but also to the information required, which may be considered of a private nature. These have to be overcome both by adopting a style of approach which will not antagonize the respondents. The question, 'How old are you?' may well bring an astute way. To mention just one of numerous possible examples, it may be required, among other things, to elicit the ages of respondents. The question, 'How old are you?' may well being an unhelpful reply, but, if half-way through the questionnaire, the year of birth of each member of the household is requested in conjunction with other routine facts, the information may be gained without difficulty. Usually it is preferable (and quicker) to present and fill in the questionnaire in person. Discussion of details such as these are given in the books noted above.

Amenity and social survey – Questionnaire sample

1 Place of birth................................sex........married/single
2 Have you always lived in this town? (name of town)
3 If not where did you live previously?
4 How long do you predict that you will stay?
5 What were your reasons for moving? Tick as relevant:
  (a) attraction of town
  (b) attraction of area
  (c) particular job was offered
  (d) good employment opportunities in general
  (e) retirement
  (f) to be near relatives
  (g) moved with parents

(h) housing situation better
(i) other reasons (specify)

6 What attracted you to this particular residence? Tick as relevant:
(a) more room
(b) previous accommodation too large
(c) good value for money/cheaper
(d) property likely to appreciate in value
(e) house well designed
(f) pleasant situation
(g) reputation of neighbourhood (e.g. friendly)
(h) to be near relatives
(i) near work
(j) near shops
(k) near school(s)
(l) other reasons (specify)

7 Employment of household

	Members of household 1	2	3	4	5	etc.
Residents in household (record relationship to head of household)						
Year of birth						
How employed (full-time job, part-time job, specifying in each case, housewife, student, etc.)						
Where employed						
How long does it take you to travel to employment?						
What means of transport is used?						

8 To housewife if not in outside employment:

(a) If the opportunity arose would you go out to work? Yes/No/Undecided

(b) Would the provision of a whole-day nursery school or créche help you in this? Yes/No/Undecided

(c) Would you be prepared to pay for this service? Yes/No/Undecided

(d) How much would you be prepared to pay each week?
£1.50 £2.00 £2.50 £3.00 £3.50 more than £3.50

9 On how many days in a month do you shop

(a) at the local shops?
(b) in the town centre?
(c) elsewhere? (specify)

10 Do you consider shopping facilities to be adequate? Yes/No

What additional shops would you like to see in the town centre?

What do you consider to be the main difficulties to shopping in the town centre?

(a) Poor bus service
(b) Lack of parking facilities
(c) Shopping centre too congested
(d) Shopping centre too spread out
(e) Car parks too far from shops
(f) Bus station/bus stops too far from shops
(g) Others (specify)

11 On how many days in a month do you

In own town Elsewhere

(a) use a public library?
(b) use a further education centre?
(c) take part in indoor sport?
(d) take part in outdoor sport?
(e) got to the cinema?
(f) go to a restaurant?
(g) go to a public house?
(h) spend evening at home?
(i) do anything else? (specify)

12 Are you satisfied with existing entertainment, social facilities? Yes/No

What additional facilities do you think should be provided in the town

(a) for children?
(b) for the youth?
(c) for adults?

The land use map

It is recommended that field research is directed towards the testing of specific hypotheses. However, by way of reconnaissance, land use and other surveys of a general descriptive nature may be undertaken in order that the student may become familiar with the varied components of the town and their arrangement. It will be appreciated that once comprehensive general surveys have been completed, raw data will be available for use in a variety of specific research projects.

Land use mapping requires the use of large-scale base maps on to which the required information can be directly recorded.

Numerical/letter notations are given for recording:

(a) number of floors of building
(b) age of buildings
(c) land use function

The notations for each of these three characteristics can be inserted on each plot so giving data from which a number of separate distributional maps may be subsequently drawn. The reconnaissance mapping team may of course be accompanied by an interview/questionnaire team collecting inside information about employment, number of residents and any other social, economic and demographic data which may be required, as indicated, for example, in the questionnaire.

(a) *Number of floors*	including cellars – indicated by numbers	
(b) *Age of building*	pre-Georgian (before 1700)	PG
	Georgian (1700–1840)	G
	early and mid-Victorian (1841–1875)	EV
	late Victorian–Edwardian (1876–1918)	LV
	inter-war (1918–1945)	IW
	post-war (1946–1960)	PW
	recent (1961–present day)	R
(c) *Function*	M manufacturing industry	

S storage and warehousing
T transport
R retail
O offices, commercial and professional
P public buildings, places of assembly, social services
X open space
D dwellings

Detail of Function:

M1 textiles	M5 vehicles	M9 paper printing
M2 clothing	M6 metals	M10 wood working
M3 engineering	M7 food, drink	M11 plastics
M4 electrical	M8 chemicals	etc. as required
O1 banks	O3 legal	O5 estate agents
O2 insurance	O4 building societies	etc. as required
S1 warehouses	S3 coal, timber yards	S4 petrol depots, garages
S2 builders yards		etc. as required
T1 bus station	T3 wharves, docks	T5 airports
T2 railway station	T4 car parks	etc. as required
P1 central government offices	P3 places of worship	P6 cinemas, theatres
P2 local government offices	P4 educational	etc. as required
	P5 hospitals	
R1 specialized shop (local)	R3 general dealer	R6 market (street)
R2 specialized shop (chain)	R4 department store	R7 catering (hotels)
	R5 market (indoor)	etc. as required
X1 parks etc.	X4 cemeteries	X6 market gardens, nurseries
X2 sports grounds	X5 allotments	etc. as required.
X3 sewage works generators		

N.B. The function refers to the ground-floor function. A section survey of multiple land use may be drawn to show the different uses of different floors, particularly valuable in the Central Business District where pressure for floor space usually gives rise to some degree of multiple land use, e.g. offices and storerooms above shops.

Simplifying the urban picture: dividing the town into general zones

A complete, detailed land use survey with the information recorded as suggested will produce a map which will probably be complicated and far from clear.

The task now is to make sense of the whole picture by dividing it into broad zones which clearly stand out, with a view perhaps to establishing whether the town or city has a concentric, sector or multiple nuclei pattern. (See section 3, pages 20–3.)

It is assumed that residential areas and industrial areas will be relatively simple to map. But the Central Business District and the Inner Zone of Transition are rather less obvious. The following key will assist in the task:

CBD components

(a) large department stores
(b) Town Hall or Civic Centre
(c) Cathedral or main church
(d) Head Post Office
(e) Central Public Library
(f) Head offices (of banks, building societies and insurance companies)
(g) supermarkets
(h) specialized national chain stores (Boots, etc.)
(i) any building of four storeys and over
(j) purpose-built hotels
(k) restaurants
(l) theatres and cinemas
(m) historic buildings

Inner zone components

(a) storage buildings (warehouses etc.)
(b) garages, car showrooms
(c) residential buildings, schools and churches **which have been converted to other uses**
(d) transport depots
(e) servicing workshops
(f) second-hand shops
(g) builders materials and do-it-yourself shops
(h) derelict property
(i) Trade Union or Labour Club
(j) older industrial premises
(k) boarding houses

N.B. There will inevitably be some overlap between the two zones, as there will be between other zones: the boundaries must therefore be generalized.

The CBD as defined by these criteria includes some buildings which it may be argued are not essentially part of the CBD in as far as they are not concerned with serving the town/city and its sphere of influence.

Practical assignments: reconnaissance surveys and hypotheses to be tested
The hypotheses given in this section provide a number of general ideas about the urban environment which may profitably be discussed as a preliminary to research work.

The quantity of research work suggested is far more than could be undertaken by one person. The best results will probably be obtained by carefully planned group work, though even then some selection will probably be necessary.

Urban components: land use in the town
Reconnaissance surveys

1 Mapping urban zones: having surveyed the town as described in the sub-section, *The land use map*, page 91, draw bounderies round and shade areas in which certain types of land use predominate (i.e. delimit functional zones), showing in particular (a) industrial zones; (i) old, (ii) new. (b) commercial zones; (i) central shopping area, (ii) outlying 'neighbourhood' centres. (c) the professional and administrative zone. (d) residential zones; (i) middle class, (ii) working class.
2 Delimit the Central Business District and Inner Zone of Transition as indicated on page 93.
3 Map any changes in land use that have occurred over a given period (e.g. the last twenty years) indicating the major zones where this has taken place and the nature of the change.
4 Map the ages of buildings in the town, drawing generalized zones of buildings constructed within given periods.
5 Draw maps to show: (a) building heights and (b) variations in land value throughout the town.

Hypotheses to be tested

1 That the most accessible locations in the Central Business District will be indicated by the intensity of pedestrian (potential customer) flow.
2 That because of more intensive competition for the most accessible (and potentially profitable) sites in the town, land values will be highest in the central areas.

3 That areas of highest land value will correspond to areas of maximum pedestrian density and that both will decline with increasing distance from the centre.
4 That within the general pattern of decreasing land value with increasing distance from the centre, there is a pattern of land value related (a) to minor nuclei of commercial and business activity (b) to lines of accessibility.
5 That because of the high demand for land in the most accessible areas the heights of buildings will be greater to provide more floor space.
6 That the pattern of land use in the town will indicate the ability of the land user to compete for the most accessible sites which will reflect his ability to make the most intensive use of the land, and that extensive land-consuming functions (e.g. manufacturing industry, warehouses, garages, showrooms) will seek peripheral locations to take advantage of lower land costs.
7 That certain types of shops selling shoppers goods (e.g. clothes, etc., for which customers 'shop around', i.e. make comparisons) will cluster in localized areas in accessible locations in the Central Business District.
8 That shops selling convenience goods which are required from day to day will seek locations within residential areas.
9 That professional services (legal, financial, etc.) will cluster together in order to make business contacts easier.

Urban population

Reconnaissance surveys. Information may be derived from statistics published by municipal agencies (a) for the town as a whole, e.g. population, occupational structure, and (b) for each ward in the town; ward boundaries are shown on 1 : 2500 maps which also indicate individual properties.

1 Establish and show in map form (a) total populations (b) number of families (c) rates of population increase (d) age structures (e) sex ratios (f) population densities per acre and per room (g) birth rates (h) death rates; (i) for a number of towns of different type, (such as those selected for question 6 on the Urban field and, in addition, a dormitory town, a New Town, a resort, an old-established industrial town) (ii) for wards within a town.

2 Show in map form, night-time and day-time population densities, and the consequent daily population flow.
3 Show in map form, the distribution of inhabitants of different racial and national origin.
4 Show in map or graph form: (a) the origins of the town's inhabitants and the reasons which prompted in-migration to the town (b) patterns of migration within the town or city.
5 Employment structure of the town: what do people in the town do? Find out from municipal records the numbers and percentages of those employed in different occupations.
6 Employment structure: what are the town's specialities and how does its employment structure differ from the national average? Work out the **location coefficients**. These are easily calculated – being the percentage employed in a given occupation in the town divided by the percentage employed in that occupation in the country as a whole; (where the two percentages are the same, the location coefficient is 1.0).
7 Portray the population structure of the whole town in the form of a population pyramid which shows the numbers of each sex in each age group. Draw population pyramids also for (a) wards within the town (b) other towns (as suggested above in 1i) (c) the same town at different periods of time (e.g. in 1900; in the inter-war period; immediately post-war).
8 Draw graphs to portray the data collected for Exercise 1 and establish the closeness of the relationships between the sets of data shown.

Hypotheses to be tested
1 That the most rapid rates of population growth have been in towns of medium size.
2 That urban population growth is closely related to growth of manufacturing industry.
3 That there is an inverse relationship between quantity of in-migration to the town and distance of migration.
4 That within towns, population growth is at its maximum in the outer suburbs and the rural–urban fringe as a result of outward migration from the inner urban areas.
5 That population density, mortality and fertility within the town show marked variations and show a close relationship to residential type.

6 That frequency of change of residence is related primarily to (a) change in the family/life cycle stage and (b) change in occupation.
7 That mobility – as measured by frequency of migration – bears a close relationship to social class (as indicated by residential type).

Urban functions and urban fields

Reconnaissance surveys

1 Mapping the urban field: how far does the influence of the town extend? Select a base map of suitable scale and plot on it: (a) where regular shoppers in the town's Central Business District have travelled from (b) the delivery areas of a number of large shops (e.g. furniture shops) (c) the places from which employees at a large office block or factory travel to work (d) the circulation area of a newspaper published in the town (e) the area served by any government office in the town, (e.g. employment exchange) (f) places smaller than the town which are linked to it by regular bus services. A line which incorporates all the points plotted in each case will indicate the town's sphere of influence for that service.
2 Establish: (a) the area (b) average radius in distance and time (c) shape and (d) population of the urban field. Distinguish between the population of the town and the population of the area outside the town.
3 How often do inhabitants of the urban field go outside it for any services; where to and what for?
4 The focus of the urban field is the town's Central Business District (the central place of the whole **field** not just of the town). What proportion of all shopping expeditions of (a) suburban dwellers (b) inhabitants of surrounding villages, are made to the Central Business District?
5 How important is the town as a service centre? The importance of a town, which will give an indication of its 'status' and rank in the urban hierarchy, can be measured by allocating points for each example of a selection of functions, which can be added to give a single total figure which can be termed '**functional importance index**'. (The score for each function gives an approximate guide to its importance: high-order functions which have high-threshold populations score more highly.)
Criteria scoring 10 points for each example recorded:

 cathedral, theatre (for plays or concerts) art gallery, daily paper, league football team, advertising agency, purpose-built dance hall, purpose-built hotel.

Criteria scoring 5 points for each example recorded: restaurant with table service, specialist book shop, cinema, institution of full-time further education (18+) other than a university, hospital, Marks and Spencer, specialist furniture shop, Woolworths, national chain food supermarket (Tesco, Sainsbury, Waitrose, etc.), department store, music shop (selling instruments).

Criteria scoring 3 points for each example recorded: national chain specialized shop (Boots, W. H. Smith, Burtons, etc.), estate agent, specialist sports shop, solicitor's office, building society, insurance company, weekly paper, music shop (selling records).

Criteria scoring 1 point for each example recorded: stationer, bank, shoe shop, fresh fish shop, men's clothing shop, secondary school, hairdresser, chemist.

This exercise is *only* useful as a comparative survey of a number of different towns, preferably of different sizes.

6 Comparative analysis

Work out areas of urban fields, populations of urban fields (indicating the separate populations of town and field) and functional importance indices for: (1) your own town (b) neighbouring towns of similar size (c) any towns that lie within (i.e. are 'nested' in) the urban field of the town surveyed (d) a large town or city within whose urban field the town you have surveyed appears to lie (e.g. the county town, if not already surveyed).

N.B. The extensive urban field of a large town is an urban field only for certain high-order functions.

Hypotheses to be tested

1 That the range and quality of urban functions varies in direct proportion to the population of the town and its sphere of influence.
2 That in all cases there is a constant ratio between the population of the town and the population of its sphere of influence.
3 That towns of high rank dispense high-order functions to extensive fields of influence (in addition to low-order functions

for local spheres of influence) whereas towns of low rank dispense only low-order functions to smaller fields of influence.

4 That people travel only short distances for low-order convenience goods required frequently, but long distances for high-order specialized goods required less frequently.

5 That the strength of influence of a town, rather than diminishing with distance from it, extends equally over the whole area of its sphere of influence as measured by the population's dependence on the town centre for goods and services, and ends abruptly at the perimeter.

Social aspects and housing in urban areas

Reconnaissance surveys. Information on these areas of enquiry may, as in the case of population data, be derived from municipal authorities as well as personal research in the field.

1 (a) What is the nature of the urban stock of living accommodation? Plot on a large-scale map of the town, the distribution of detached houses used by individual families, semi-detached houses, terraced houses, purpose-built flats, flats converted from residences once used by individual families, institutions which accommodate people on a permanent or semi-permanent basis, (e.g. old people's homes, homes for incurable patients) (b) Show in the form of a bar graph the numbers living in each type of accommodation.

2 What is the condition of the urban stock of living accommodation? Show in map form the distribution of houses classified qualitatively by numbers of rooms (per inhabitant if available), washing and sanitary facilities and the proportions living in houses classified as suffering from blight.

3 How old is the urban housing stock? Show the proportion of the town's population living in property of different ages.

4 How is the housing stock owned? What proportion of the town's inhabitants own their own property, rent accommodation privately, rent housing from the Council?

5 Where do people want to live? Design a **popularity index** by comparing house prices and trends, and by questionnaire.

6 How clearly marked are social differences in the town? Map the variations which exist in factors such as income, occupation, education, voting habits, participation in communal activities, attitudes to the nature of urban life.

7 How do people see their situation in relation to others? Examine people's views on such matters as neighbourliness, dropping in, privacy, loneliness, social facilities, the council, the police, the ideal neighbour, new development schemes, landlords, coloured immigrants, the advantages and disadvantages of flat dwelling, open-plan estates.

8 How are social problems dealt with? What provision is made: (a) publicly (b) privately, for the health and welfare of all sections of the population, particularly the young, the old, the ill and poor?

9 Draw a map to show the distribution of social and housing problem areas and the incidence of crime in the town.

10 What differences exist between ways of life and attitudes in the country and in the town? Examine the attitudes of a sample of country dwellers and a sample of urban dwellers on a range of issues which appear to you important, e.g. (a) the advantages of living in the country (b) the advantages of living in the town (c) matters suggested for investigation in Question 7.

Hypotheses to be tested

1 That housing densities (excluding flats) have been progressively reduced over the last century, but unevenly as regards different social groups.

2 That there is a close correlation between housing quality and the incidence of social problems.

3 That poverty is peculiar to the inner zone of urban transition.

4 That social areas are identifiable, representing a tendency for people with similar social characteristics to group together.

5 That geographical mobility is closely related to social class and shows significant variations between the inhabitants of different housing classes.

6 That the incidence of low income, poor housing, occupational insecurity and low educational attainment are localized and occur in inner urban areas.

7 That the localization of immigrant communities results from: (a) inability to compete in the housing market (b) social discrimination (c) desire for cheap accommodation (d) desire to be among one's own national group.

8 That in general, where people live is determined by: (a) stage in the family life cycle (b) financial resources (c) social

aspiration (d) class identity and solidarity (e) proximity to desirable amenities, in that order.

9 That the strength of social ties with one's own kin: (a) has declined with the passage of time (b) differs significantly between classes.

10 That frequency of social interaction declines with advancing years.

11 That people seek privacy before neighbourliness in their choice of residence.

12 That in terms of social characteristics and attitude, rural–urban differences are not now signficant.

Transport, amenities and environmental quality

Reconnaissance surveys

1 Traffic density: draw a map to show the density of traffic flow on selected roads in the town indicating vehicles per hour by lines of appropriate width. Show where traffic hold-ups occur most frequently.

2 Traffic speed: draw a map to show average traffic speeds on selected roads in the town indicating speeds by lines whose width is inversely proportional to speed.

3 Plot the distribution of traffic accidents in the town.

4 Plot the distribution of car parks and their relationships to shopping areas.

5 Show in map and graph form, the proportions of people travelling by private car, bus, cycle, rail: (a) for travel within the town (b) for travel into and out of the town.

6 Work out the proportion of all land in the town devoted to transport and movement (include roads, pavements, railway stations, etc.).

7 Future plans: collect maps to show proposed changes in the transport system.

8 *Service amenity indices:* work out service amenity indices by adding the time taken (from a selection of houses) to reach a range of important services: post office, newsagent, food shop, doctor, bus stop, railway station, public house.

9 *Recreational amenity indices:* work out the time taken for separate visits to open green space, moorland, mountains, woodland, seaside, lake, river (for fishing or boating).

10 What is the range of social and recreational provision in the town for: (a) children (b) the middle-aged (c) the old?

11 Who is responsible in the town for pollution and noise control; preservation of historic buildings; the renovation and up-keep of property; street litter? Is there a Civic Trust in your town? If so, what does it do?

12 Land use conflict: where is this most obvious? (e.g. green belt, *v.* industry, road expansion *v.* residential quality.)

13 Do people care about the ugliness or beauty of the environment? Devise a questionnaire to establish: (a) attitudes to the visual quality of the environment (b) what people are prepared to do about it.

Hypotheses to be tested

1 That traffic density is inversely proportional to traffic speed.
2 That frequency of traffic accidents is closely related to traffic density.
3 That improved neighbourhood planning is shown by close relationship between age of property and amenity indices.
4 That most people put environmental appearance as a low priority.

14 Books and other sources of information

Many good alternatives may be more readily obtained from local school and public libraries or purchased cheaply in paperback editions.

The prices can only be taken as an approximate guide to the cheapest editions.

Useful Dewey Decimal numbers are: 300–399 (Social Sciences), 711 and 720 (Architecture and Town Planning), 910 and 914 (Geography), 942 (History).

Geography of Settlements, F. S. HUDSON, Macdonald and Evans, 1970, £2.00

Urban Geography, J. H. JOHNSON, Pergamon, 1967, £1.20

Towns and Cities, E. JONES, OUP, 1966, £0.40

These three books provide together a full introductory coverage of all aspects of urban study and may be viewed as 'first phase' reference works. Though geographically oriented they include excellent sections on the origins and historical development of towns and cities.

Living Space, R. BOOTH, Intertext, 1968, £1.00

Deals with a wide range of social and planning issues including urban regional planning, the planning and design of buildings, traffic, urban renewal and housing.

Places for People: Cities and Regions, J. BRADLEY, Intertext, 1968, £0.40

Includes suggestions for project work and examples of projects undertaken by students.

Architecture: City Sense, T. CROSBY, Studio Vista, 1965, £0.75

A challenging introduction to town planning from the architect's point of view.

Town Study: a Sample Urban Geography, K. DRISCOLL, Philip, 1971, £1.50

A case study in depth of an imaginary British town of medium size; an excellent reference book for the work suggested in the Practical Assignments section.

Settlement Patterns, J. EVERSON and B. FITZGERALD, Longmans, 1969, £1.00

Introduces the new quantitative approach to geographical settlement study.

Communities in Britain, R. FRANKENBERG, Penguin, 1970, £0.40

Sociological case studies of different types of urban and rural communities.

Knowhow, R. HARE, Intertext, 1970, £1.75

Deals with the practical and mechanical details of field research.

Patterns of Urban Life, R. PAHL, Longmans, 1970, £0.75

An introduction to urban sociology.

Geographical

Metropolis on the Move: Geographers look at Urban Sprawl, ed. J. GOTTMAN and A. HARPER, Wiley, 1967

Image of the City, K. LYNCH, MIT, 1960

The Geography of Towns, A. E. SMAILES, Hutchinson, 1966

Methods of enquiry and presentation

Quantitative Geography, S. M. BARKER and W. V. TIDSWELL, UTP, 1970

Field Work in Urban Geography. K. BRIGGS, Oliver and Boyd, 1970

Town Study Companion, G. B. G. BULL, Hulton, 1969

Local Geography in Towns, J. HADDON, Philip, 1971

Historical

Charles Booth's London, ed. A. FRIED and R. ELMAN, Penguin, 1971

Victorian Cities, A. BRIGGS, Penguin, 1968

The Town Labourer (1760–1842), J. L. and B. HAMMOND, Longmans, 1966

The City in History, L. MUMFORD, Penguin, 1966

Social

Shape of the Community, S. CHERMAYEFF and T. ALEXANDER, Penguin, 1971

Poverty: the Forgotten Englishman, K. COATES and R. SILBURN, Penguin, 1970

Count Us In: a Community Service Handbook, M. DICKSON, Dobson, 1970

The Death and Life of Great American Cities, J. JACOBS, Penguin, 1972

Patterns of Infant Care in an Urban Community, J. and E. NEWSON, Penguin, 1971

Race, Community and Conflict, J. A. REX and R. MOORE, OUP, 1969

Family and Kinship in East London, M. YOUNG and P. WILLMOTT, Penguin, 1969

Transport

Traffic in Towns, C. BUCHANAN, Penguin, 1969

New Movement in Cities, B. RICHARDS, Studio Vista, 1966
Homes, Towns and Traffic, J. TETLOW and A. GOSS, Faber, 1970
Motorways in the Urban Environment, LLEWELLYN DAVIES etc., British Road Federation, 1971

Planning

Planning for People, M. BROADY, NCCS, 1969
London 2000, P. HALL, Faber, 1969
Garden Cities of Tomorrow, E. HOWARD, Faber, 1945
New Towns Come of Age, Town and Country Planning Special Issue Jan, 1968

Further sources of information

Periodicals and journals

1 The Geographical Magazine, 128 Long Acre, London WC2 E9QH. Issues particularly during the period since 1970 have included excellent articles on towns and cities, past, present and future.
2 Town and Country Planning, 28 King Street, Covent Garden, London WC2. Excellent coverage of today's problems and means of solving them.
3 The Journal of the Town Planning Institute, 26 Portland Place, London W1.
4 Official Architecture and Planning
5 Geography
6 Geographical Journal
7 Urban studies
8 New Society

National Press

9 Observer
10 Sunday Times
11 Times
12 Guardian
13 New Society

(9–13) news and special features will furnish copious source material for projects

Other sources

Commission for New Towns, Glen House, Stag Place, London SW1
British Road Federation, 26 Manchester Square, London W1
Civic Trust, Walter House, Bedford Street, London WC2
Shelter, 86 Strand, London WC2

 Regional Studies Group, University of Strathclyde, Glasgow C1
Plan Europe 2000, c/o European Cultural Foundation, Jan van Goyenkade 5, Amsterdam, Netherlands
H M S O, 49 High Holborn, London WC1

Planning departments

1 Glasgow, Sheffield, Cardiff, Liverpool, Newcastle, Birmingham, Greater London, Manchester (particularly for housing)
2 Leicester, Greater London (particularly for traffic)
3 Plymouth, Coventry, Sheffield, Birmingham, Manchester, Liverpool, Durham, Greater London (particularly for town centre renewal)
4 York, Chester, King's Lynn, Bath, Sandwich (particularly for preservation and renewal)
5 Stevenage, Crawley, Harlow, Welwyn Garden City, Basildon, Cumbernauld, Cwmbran, Corby, Dawley, etc. (particularly for New Town information)
6 Universities of Durham, Reading, Lancaster, Glasgow, Strathclyde (only if detailed information for a study in depth is required)

Local sources

Library, museum, city archives, local Planning Department, local Government Offices, Health and Social Service Department, Rating office, Borough Engineer (traffic), estate agent, architect, local newspaper offices, school and college (geography and history) staff, party political branch headquarters.

15 Key concepts: a test

These are factual questions, and of the five suggested answers one is right and all the others are wrong. Put a circle round the letter next to the one which you think is the correct answer. The test is based on concepts introduced in the book.

1 Central Place

(a) the Central Business District of a town
(b) the point of maximum pedestrian and traffic flow in a city; usually associated with a high level of commercial activity
(c) Colin Buchanan's London address.
(d) a town or settlement which acts as a centre for providing services to a surrounding area
(e) the square in the centre of a settlement often occupied by a market.

2 Convenience Goods

(a) goods bought nearly every day from a nearby shop
(b) goods that are easily obtainable
(c) goods which fall into the category of 'labour-saving devices'
(d) goods only available in an accessible central urban area
(e) family planning devices dispensed through the N.H.S.

3 Expanding Town

(a) a town absorbed into a city by the process of urban sprawl
(b) a town which has received 'overspill' population from a large conurbation and is expanding and developing
(c) a New Town established to accommodate the excess population of an expanding city
(d) planner's jargon for a town which is getting smaller
(e) a small town whose population is increasing at a rate above a given level.

4 Friction of Distance

(a) a term referring to the effects of heated arguments between rush-hour travellers and railway officials
(b) the limited room for expansion of firms in old congested areas within towns
(c) the irritability and fatigue resulting from increasing lengths of journeys to work
(d) the microclimatic effects of increased temperature in urban environments
(e) a term used to describe the phenomenon that the further towns are away from each other, the less is the frequency of inter-relationship between them.

5 **Functional Zone**

(a) an area in which a particular functional type of land use predominates
(b) an area in which there is a wide variety of urban functions
(c) a residential zone with high fertility rates
(d) part of a town with a large number of public conveniences
(e) an area within a town's administrative boundary allocated for industrial expansion.

6 **Ghetto**

(a) part of a town designed for people who want to 'get away from it all'
(b) a covered, traffic-free pedestrian shopping precinct
(c) an area of a town used for semi-permanent accommodation by squatters
(d) a slum area with a high proportion of deprived occupants often of one race or religion
(e) an area which is in the process of getting rid of its traffic problems.

7 **Hinterland**

(a) the area served by a town
(b) a backward area
(c) the area surrounding a town which contains settlements linked by bus services
(d) the area from within which a town's water supply is taken
(e) the land between or on the boundary between urban fields.

8 **New Town**

(a) a town build after the 1946 New Towns Act
(b) a town which has undergone a comprehensive programme of redevelopment
(c) a satellite town which has agreed to accommodate the excess population of a large conurbation
(d) a Liverpool suburb with a high crime rate
(e) a planned town developed under a Development Corporation.

9 **Node**

(a) a central and accessible focus of activity

(b) the central point of an urban transport system
(c) a road with 'No Entry' signs at both ends
(d) a point of concentration of political and administrative power
(e) the attraction of rural population to towns.

10 Overspill

(a) a condition that frequently occurs at accident black spots
(b) the excess population of a large city for whom accommodation is provided in surrounding settlements
(c) the condition that exists immediately prior to large-scale emigration from a city
(d) the population of an area about to undergo urban renewal
(e) the population of an expanding city who find themselves living beyond the city boundaries.

11 Principle of Least Effort

(a) the principle by which towns are allowed to develop and expand naturally without the interference of planners
(b) the principle applied to the planning of recreational facilities in urban areas
(c) the principle by which human settlements are arranged in the most efficient way
(d) the principle which governs the planning of new primary schools in towns
(e) the principle by which the majority of the population do the least possible amount of work.

12 Social Area

(a) an area of the town whose inhabitants are mostly Labour Party voters
(b) an area of the town whose inhabitants are socially similar
(c) an area of the town where there is a high degree of participation in social and cultural activities
(d) a recreation ground in a city
(e) an area of the town with a preponderance of inhabitants of middle and upper-middle class.

13 Threshold Population

(a) the population needed to support a particular service or facility
(b) the upward limit of population growth in an expansion plan
(c) those people in a town who spend much of the day standing by the front door chatting to people in the street
(d) the population which inhabits a town's commercial catchment area
(e) the population of the outer suburbs.

14 Twilight Zone

(a) an area of a town where the work force operates between the hours of 4 pm and 8 pm.
(b) the zone of a town where cinemas, theatres and other forms of entertainment are located
(c) the area of the town where land uses and functions are obsolete and in decay
(d) an area which has undergone urban renewal
(e) a city's Western suburbs.

15 Urban Leapfrog

(a) the continuous process of migration from inner city areas to the edge of the built-up area
(b) a game for ambitious city business men
(c) a process of intra-urban movement associated with the 'Green Belts'
(d) an increasingly common method of queue jumping on crowded pavements, resulting from increasing traffic congestion in inner-urban areas
(e) the process of socio-geographic mobility in urban areas.

16

Facts test

These are factual questions and of the five suggested answers one is right and all the others are wrong. Put a circle round the letter next to the one which you think is the correct answer. (In cases where, for one reason or another, exact correctness may be disputed, the wrong answers are significantly wide of the mark.) The questions are derived from information given in the book.

1 (i) The number of million cities in the world in 1850 was

(a) 2
(b) 10
(c) 60
(d) 70
(e) 80

(ii) The number of million cities in the world in 1961 was

(a) 40
(b) 60
(c) 80
(d) 100
(e) 250

2 (i) The most urbanized continent is

(a) Europe
(b) N. America
(c) Oceania
(d) Africa
(e) S. America

(ii) The least urbanized continent is

(a) N. America
(b) Oceania
(c) Africa
(d) S. America
(e) Asia

3 (i) The population of Rome during the Roman Empire reached an estimated maximum of

(a) 5 000
(b) 25 000
(c) 50 000
(d) 200 000
(e) 1 000 000

(ii) The population of Roman London reached an estimated maximum of

(a) 5 000
(b) 30 000
(c) 50 000
(d) 95 000
(e) 500 000

4 (i) London: the population density (in population per square mile) recommended for inner boroughs in 1951 ranged from
(a) 3 500 to 10 000
(b) 14 500 to 29 000
(c) 31 500 to 56 000
(d) 62 000 to 85 000
(e) 310 000 to 410 000

(ii) London: the population of 'The City' changed between 1851 and 1961 from
(a) 128 000 to 8 800 000
(b) 48 000 to 630 000
(c) 128 000 to 129 000
(d) 48 000 to 38 000
(e) 128 000 to 5 000

5 (i) The highest extremes of population density are to be found in
(a) Hong Kong
(b) Paris
(c) London
(d) Welwyn Garden City
(e) Manhattan

(ii) The extreme recorded density (in population per square mile) is
(a) 5 000
(b) 50 000
(c) 500 000
(d) 5 000 000
(e) 50 000 000

6 (i) The normal working-hour population of the Empire State Building is
(a) 25
(b) 250
(c) 2 500
(d) 25 000
(e) 250 000

(ii) A technically feasible Tower City for a population of 300 000 has been designed to a height as great as
(a) 300 m.
(b) 1 000 m.
(c) 1 500 m.
(d) 3 750 m.
(e) 10 000 m.

7 (i) In 1963 an advertisement designed to encourage office dispersal made the observation that the floor space for a waste paper basket (of average dimensions) in Central London could well cost in rental and rates per year
(a) £0.20
(b) £0.60
(c) £9.80
(d) £1.30
(e) £2.65

(ii) According to the Guinness Book of Records, the most expensive land in the world in 1972 (in the City of London) was valued at
(a) £100 per square metre
(b) £500 per square metre
(c) £1250 per square metre
(d) £1850 per square metre
(e) £3000 per square metre

8 (i) The average frequency of visits to furniture shops (with intent to purchase) in weeks per visit is
(a) 1
(b) 5
(c) 11
(d) 32
(e) 52

(ii) The average frequency of visits to chemist shops in weeks per visit is
(a) 0·5
(b) 1·0
(c) 3·2
(d) 6·7
(e) 11·0

9 (i) The minimum population required to support a Boots Chemist is
(a) 1 000
(b) 10 000
(c) 30 000
(d) 100 000
(e) 500 000

(ii) The minimum population required to support a John Lewis department store is
(a) 10 000
(b) 30 000
(c) 100 000
(d) 500 000
(e) 1 000 000

10 (i) The increase in number of vehicles per head in Britain between 1910 and 1950 was from
(a) ·01 to ·03
(b) ·01 to ·09
(c) ·001 to ·13
(d) ·03 to ·18
(e) ·3 to 1·8

(ii) The increase in number of vehicles per head in Britain between 1970 and 2000 (estimated) will be
(a) ·17 to ·25
(b) ·17 to 2·5
(c) ·33 to ·55
(d) ·21 to 1·00
(e) ·75 to 2·10

11 (i) The 1970 estimated annual loss directly resulting from traffic delays in England and Wales was
(a) £5 000
(b) £50 000
(c) £500 000
(d) £5 000 000
(e) £500 000 000

(ii) The average vehicle journey speed in mph in Central London changed between 1909 and 1958 from
(a) 2·5 to 29·9
(b) 10·1 to 28·2
(c) 13·7 to 23·7
(d) 10·1 to 12·0
(e) 11·4 to 10·3

12 (i) The percentage of homes in England and Wales without indoor lavatories in 1969 was
(a) 15·9
(b) 21·3
(c) 40·2
(d) 52·3
(e) 58·4

(ii) The percentage of homes without any lavatory in 1969 was
(a) 2·6
(b) 4·6
(c) 10·1
(d) 15·2
(e) 21·3

13 (i) Under whose auspices was New Lanark built?
(a) Titus Salt
(b) Sir Alec Douglas Home
(c) George Cadbury
(d) Robert Owen
(e) Charles Clore

(ii) The name of which of the following urban planners has been given to traffic-free-residential arrangement?
(a) Howard
(b) Geddes
(c) Olmsted
(d) Buchanan
(e) Radburn

14 (i) The first New Towns Act was passed in
(a) 1923
(b) 1936
(c) 1963
(d) 1946
(e) 1496

(ii) The first New Town was
(a) Telford
(b) Grimsby
(c) Harlow
(d) Stevenage
(e) Newton Aycliffe

15 (i) The first Garden City was
(a) Hampstead
(b) Letchworth
(c) Gorbals
(d) Babylon
(e) Solihull

(ii) Where was/is the first planned New City in Britain?
(a) Plymouth
(b) Newhaven
(c) Bracknell
(d) Birmingham
(e) Milton Keynes

17

Test answers

Concepts	
1	d
2	a
3	b
4	e
5	a
6	d
7	a
8	e
9	a
10	b
11	c
12	b
13	a
14	c
15	a

Facts	(i)	(ii)
1	(i) a	(ii) c
2	(i) c	(ii) c
3	(i) d	(ii) b
4	(i) b	(ii) e
5	(i) a	(ii) c
6	(i) d	(ii) d
7	(i) e	(ii) e
8	(i) d	(ii) b
9	(i) b	(ii) c
10	(i) b	(ii) c
11	(i) e	(ii) e
12	(i) a	(ii) a
13	(i) d	(ii) e
14	(i) d	(ii) d
15	(i) b	(ii) e

Acknowledgements

Permission to reproduce illustrations is gratefully acknowledged to the following:

The Institute of British Geographers for Fig. 12 (from 'Some Reflections on the Geographical description and analysis of towns' by A. E. Smailes, from *Transactions* 1955); the Ordnance Survey, for Figs 18 and 25; the Cumbernauld Development Corporation for Figs 19 and 20; British Rail for Fig. 26; Ian Mullely for Figs 27 and 28; Aerofilms Ltd for Figs 29, 30, 32, 33 and 34; the Radio Times Hulton Picture Library for Fig. 31; Aerofilms Ltd and Ewing Galloway for Fig. 35; the Allen Clarke Research Institute, The Plessey Company, for a Cambridge 'Stereoscan' micrograph, Fig. 36.

www.ingramcontent.com/pod-product-compliance
Lightning Source LLC
LaVergne TN
LVHW021113300926
849160LV00009B/718
9780415851855